W9-BNM-756

About Island Press

Since 1984, the nonprofit organization Island Press has been stimulating, shaping, and communicating ideas that are essential for solving environmental problems worldwide. With more than 1,000 titles in print and some 30 new releases each year, we are the nation's leading publisher on environmental issues. We identify innovative thinkers and emerging trends in the environmental field. We work with world-renowned experts and authors to develop cross-disciplinary solutions to environmental challenges.

Island Press designs and executes educational campaigns in conjunction with our authors to communicate their critical messages in print, in person, and online using the latest technologies, innovative programs, and the media. Our goal is to reach targeted audiences—scientists, policymakers, environmental advocates, urban planners, the media, and concerned citizens—with information that can be used to create the framework for long-term ecological health and human well-being.

Island Press gratefully acknowledges major support of our work by The Agua Fund, The Andrew W. Mellon Foundation, The Bobolink Foundation, The Curtis and Edith Munson Foundation, Forrest C. and Frances H. Lattner Foundation, The JPB Foundation, The Kresge Foundation, The Oram Foundation, Inc., The Overbrook Foundation, The S.D. Bechtel, Jr. Foundation, The Summit Charitable Foundation, Inc., and many other generous supporters.

The opinions expressed in this book are those of the author(s) and do not necessarily reflect the views of our supporters.

Quantified

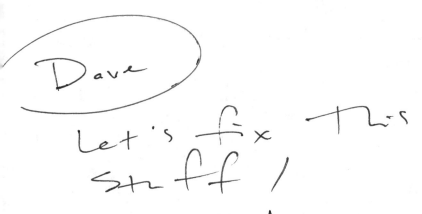

Dave

Let's fix this stuff!

Quantified

REDEFINING CONSERVATION
FOR THE NEXT ECONOMY

Joe Whitworth

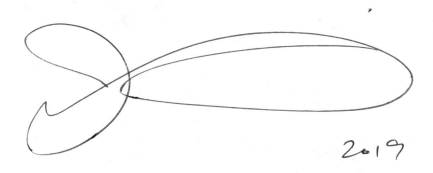

2019

ISLANDPRESS

Washington | Covelo | London

Copyright © 2015 Joe Whitworth

All rights reserved under International and Pan-American Copyright Conventions. No part of this book may be reproduced in any form or by any means without permission in writing from the publisher: Island Press, 2000 M Street, NW, Suite 650, Washington, DC 20036.

ISLAND PRESS is a trademark of the Center for Resource Economics.

Library of Congress Control Number: 2015934557

Printed on recycled, acid-free paper

Manufactured in the United States of America
10 9 8 7 6 5 4 3 2 1

Keywords: environmental markets, environmental policy, environmental lawsuits, environmental philanthropy, sustainable agriculture, water, Mississippi River, Klamath River Basin, Colorado River, Australia drought, The Freshwater Trust

For Liz, Ellie, Anna, and Henry.
And all others who build from here.

Contents

A Note to Readers

As president of The Freshwater Trust, I long ago committed to innovating beyond what was known in order to do what was needed. This led us to engage technology and create new methods to accelerate the pace and scale of restoration of freshwater ecosystems. Where some saw our commitment to experimentation and evolution as bordering on maniacal, others encouraged me to share these new tools with a broader audience by distilling them into a book. Having never written a book, I turned to Andrea Carlos, an accomplished journalist with an abiding interest in conservation, and despite having a farm remodeling project under way at the time, she agreed to help. A superb collaborator, she is a key reason why this project got done. This is how we worked together: After distilling the key elements of the book framework, I laid out the original thought line of how the economy and the environment must integrate in the face of twenty-first-century realities. Thereafter, Andrea and I figured out the right stories, research, and expert interviews needed to create the countless drafts, which we passed back and forth to hone the manuscript into its current state. Although I would not describe the work as easy, we both feel that the partnership rendered some great stuff; we hope you will agree.

Acknowledgments

I get paid to tell the truth—not necessarily an easy thing in a world that does not always want to hear it but something I feel compelled to do nevertheless. Our ability to do so depends on three basic factors: We need to want to do it, we need to know how to do it, and we need to be in a position to do it. Each of us is a composite of our experiences and relations—shaped greatly or slightly by every single interaction—and book writing is no different. Neither this work nor my perspective would be here without all the arguments, discussions, and learning graciously afforded to me throughout my life. My exceptional mom would say, rightly, that the events and people leading up to this publication are the real catalysts here, and I am forever grateful for what they have given me.

Before writing this book, I did not know what drove me. I did not have a singular, mind-blowing wilderness experience that propelled me into this work but rather a quiet and steady influence from my grandpa and dad, delivered over minutes, hours, days, and years. Neither of them will ever read this book, but both can clearly be read in me. In wrestling with this project, I gained a perspective on my relations with them that will forever guide my mind.

The people I work with provide me with more energy and insight than anyone has a right to, and I count myself fortunate to be able to work with such committed intellect. This includes the staff, top to bottom, of The Freshwater Trust since the day I walked in the door. As individuals and as a group, we remain interested not in stuff that sounds cool or merely looks good; we need it to work for real. I hope to have the privilege of working alongside many of these colleagues for years to come. Those who directly helped inform and shape this project include Caylin Barker, Brett Brownscombe, Matt Desmond, Joe Furia, Marley Gaddis, Tony Malmberg, Adrian McCarthy, Mark McCollister, Jason Miner, Gustavo Monteverde, Jim Myron, David Pilz, Karin Power, David Primozich, Erin Putnam, Nicole Spencer, Haley Walker, and Tim Wigington. The guy who gave me the final push to get it done was Alan Horton, a trusted colleague fully committed to changing the world for the better.

Because my vocation is the avocation of others, I meet folks in a unique head space: They want to bring their intellect and resources to bear on the big problems we face as a society. There have been many who helped along the way, but those who pushed me hardest, informed me best, and supported me unstintingly include Hank Ashforth, Roger Bachman, Reed Benson, Tim Boyle, Andy Bryant, Dave Chen, John Colosimo, Scott Demorest, Rocky Dixon, Matt Donegan, Gary Fish, Paul Fortino, Al Jubitz, Art Kayser, Mike Keiser, Don Krahmer, Randy Labbe, Dave Laurance, Lynn Loacker, Luis Machuca, Marty Myers, Jan Newman, Tim O'Leary, Brad Preble, Scott Sandbo, Bill Smith, Tony Trunzo, John von Schlegell, and David Willmott. I found myself returning frequently to discussions and scenarios with this group as I wrote, and I thank them for their good counsel.

Colleagues and friends both inside and near the conservation community provide a deep well of inspiration, but fixed in my mind are a group whose conversations I turned to repeatedly whenever I had difficulty navigating, which happens when you undertake a book project: Bill Bakke, Ricardo Bayon, Mike Blumm, Fred Boltz, Paul Brest, Alexandra Cousteau, David James Duncan, Marshall English, Peter Gleick, Robert Glicksman, Martin Goebel, John Goldstein, Bill Hatcher, Deb Hatcher, Kenny Helfrich, Rick Henslee, Dan Keppen, Jim Klug, Ben Koldyke, Gregg Lemkau, Ian Lombard, Patrick Maloney, Ned McCall, Nancy McKlveen, John Nordgren, Patrick O'Toole, Wendy Pabich, Dick Pedersen, Jim Prosser,

Andrew Purkey, Dan Rohlf, Jason Scott, Mary Scurlock, Susan Phinney Silver, Peter Stein, Robert Stubblefield, Gene Sykes, Dan Winterson, and James Workman. The beacons provided by this group of advocates, doers, innovators, scientists, and scholars truly kept me focused and the project on track in trying to describe the needed evolutions in accounting, agriculture, economics, and the environmental movement. Any errors, miscalculations, and other such failings are my own.

My editor, Emily Davis, was in many ways the best part of doing this project. Her accessibility and willingness to work through the daunting task of laying out a plan to redesign conservation and integrate it into a durable economy make her legendary in my mind. The most difficult thing I never knew existed—copy editing—was flawlessly tackled by Sharis Simonian. On the marketing side, Julie Marshall and Jaime Jennings made it their jobs to put this book in your hands—they not only did it, they made it look easy. This is my first book, and as publisher, Island Press nailed it.

In the end, I wrote this with not past but future generations in mind. Whenever I came to a point where I could honor those who got us this far while pointing out where we must go next, I did exactly that. But in all conflicts between what we used to need and what we need now, I intentionally sided with those who must face the complex problems ahead: you.

Introduction

The seeds of my current work were first planted by my grandfather more than 40 years ago. A corn and bean farmer along Blackbird Creek in the Missouri River Basin, he used to say that no man has the right to take more from the land than what the land itself can withstand. Over decades, he learned that if he took care of the land, it would take care of him. In a fundamental way, he understood that commerce and environmental stewardship were forever entwined, that prosperity requires both a strong economy and a healthy environment.

As a young college graduate, I entered a world that sends a quite different message. Not only are the economy and the environment seen as completely separate, but they are *at war*. The message is that you can have either a strong economy or a strong environment but not both. This view of the world has never sat well with me. And when I see evidence that both our economy and our environment are in decline, I'm reminded that Grandpa Whitworth had it right.

Yet despite my grandfather's good intentions, Blackbird Creek has since been listed in violation of the Clean Water Act.[1] In fact, its entire length suffers from agricultural runoff, including the stretch of creek that bordered my grandfather's property. Like most farmers, my grandfather intended to do right by the land. Yet he was caught up in an economy that didn't bother to connect the dots. Fertilizer was cheap. Conventional wisdom was to farm all the way down to the stream. And an ongoing need to pay off bank loans required him to keep increasing the number of bushels he produced. Unfortunately, my grandfather's experience is still the rule, not the exception, in today's world. It's just the system.

A Finite Sandbox

Like my grandfather, most people do not fully understand or connect their actions to the impact they have on the environment. Producing cheap food requires lots of fertilizer and pesticides. And because we all enjoy cheap food, our nation's rivers are literally choking from these nutrients. Take the Mississippi River, for example. The world's fourth longest river, it has become so saturated with fertilizers from agricultural runoff that every year it creates an enormous water ghost covering as many as 8,000 square miles—the size of New Jersey—where the river drains into the Gulf of Mexico. Within that dead zone, there's not enough oxygen to sustain fish or other marine life. In their place are enormous quantities of toxic algae, leaving an ugly layer of scum to shadow the depths below.

We're taking more from the land than it can withstand, and in the long term that's bad for the economy. With the world's population projected to reach 10 billion by 2050, we'll have more people to feed, clothe, house, and employ than in any time in human history. *And then we'll have to do it every year thereafter.* That means we cannot undercut the resource base from which we draw. We have a limited sandbox in which to play. Our natural resources are finite. They are the basis of our prosperity; we can't just use them up.

Yet that's exactly what we've been doing. In the last 150 years since the Industrial Revolution, we've focused almost exclusively on growing our economy, extracting whatever we need to do so at the expense of the environment. And now we're at a point where the environment is hurting—seriously hurting. Setting aside all romantic reasons for saving nature, the fact is, we cannot have a thriving economy without a resilient resource base underlying it. It just won't work. We need basic systems operating properly, and right now, they are deeply compromised as a result of humans making a living on Earth.

We're in a game of catch-up, and to restore the right balance, we must train a laser focus on achieving gains for the environment in the same way that we've obtained financial gains in the past. Simply put, we have to rebuild the health of the environment on whose services both our economy and our very existence depend. In my mind, this is not a war where we must halt the evil economy in the name of a beautiful environment as foretold by eco-warrior legend. This is an obvious imperative.

Busywork or Actual Results?

Growing up in a small town in downstate Illinois, I was also influenced by my dad, a carpenter. I spent my summers as a grade-schooler earning a dollar a day running back and forth to the truck getting the right tool for the job at hand—a great way to understand the tools and learn the trade. The way it worked was pretty simple. People called my dad when they had a problem. We would show up on the job site, check out the situation—the gutter would be broken, the roof would be leaking, the floor would have fallen through—and if Dad did his job right, the problem would be fixed by the end of the day. But if we walked off that job site and the problem wasn't fixed, we'd have to come back the next day and then the day after that until it was. Dad didn't get paid unless and until he fixed the problem.

Being exposed to job site after job site all the way through high school, I learned that when there's a problem, you fix it. Yet when I started working on water issues, I soon recognized that we weren't getting the job done. Yes, we were raising money every year. Yes, we were staying busy. Yes, we were helping. But we weren't actually fixing the problem.

When you're a carpenter, the first thing you do is size up a problem. Is it a leaky roof or a cracked foundation, and what tools do I need to fix it? Sizing up our twenty-first-century water problems, I eventually reached the conclusion that the tools we've been using aren't enough to solve the challenges we face. It's like trying to paint an entire house with a 1-inch paintbrush when what you really need is a spray gun. In the same way, the tools we're using to protect the environment aren't getting us where we need to be. Despite the hard work by a lot of uber-smart, totally dedicated, gifted people, the reality is that we're not achieving the gains we need for the environment. In fact, the environment continues to lose ground at a rapid pace.

Taking a Quantum Leap Forward

I have written this book for the same reason that I work on water issues: I want to take my grandfather's good intentions about the environment and convert them into action while still allowing people like him to earn a decent living. And I want to harness my dad's fix-it work ethic to restore our rivers and streams within my lifetime. Unless we change the course we are on, we simply won't get it done.

There's an incredible array of tools available to us, but we have yet to seize them. Instead, we remain stuck in Conservation 1.0—an unacceptable rate of innovation for smart folks living in the age of Google. Most of the major advances the environmental movement has made date back to the 1970s, when the Clean Water Act was passed and issues such as clean water and air drew national attention. True, there's been a tremendous amount of advocacy and litigation since then. But the improvements have been incremental. Perhaps we've moved on to Conservation 1.1 or 1.2, when what we really need is a quantum leap forward. The bottom line is that the environmental movement hasn't been innovative, and we need to be innovative if we're to address the complex environmental problems in front of us.

To use an example that most environmentalists hate, consider the oil and gas industry. Historically, oil was extracted by drilling vertically. Vertically, vertically, vertically—for more than a century. But over time, the work the industry could get done diminished. The oil began to dry up, and the return on investment began to decline. Rather than throwing up their hands, oil and gas executives regrouped and innovated. They looked sideways at the issue and came up with a new way of getting at the oil that involves drilling horizontally rather than vertically. No matter how you feel about the substance of this example, you cannot argue with the form. Horizontal fracking opened up a whole new oil boom.

In the same way, we need to realize that the same old environmental playbook is no longer working and find a way to open up new possibilities. We've spent the last 45 years using the same old tools and strategies, even as the return on our investment has diminished. Yet drilling down harder on our problems using these same methods isn't going to work. We've got to wake up and realize that we're not getting at the issue and that we have to move sideways, at an angle, or in some other way. In short, we need to tackle our environmental problems from a different direction, and that requires a brand-new approach.

Whatever our role working on environmental issues, we must all have a come-to-Jesus meeting with ourselves and our organizations. We must take a hard look at where we're at, admit where things aren't working, and then revamp our approaches to get the results we want. And we must innovate and measure our results to make sure our chosen path is working.

Quantified is about doing exactly that. It's about changing our approach to conservation on a fundamental level. It's about widening the focus to bring about environmental gains alongside the financial ones that have been the central emphasis of our global economy. And it's about moving past the current "let's stop more bad things from happening" mentality to achieve lasting, quantifiable improvements for the environment. Although many of the examples in this book come from my field of focus, water, the principles of quantified conservation apply to environmentalists working on any issue. They also apply to the entire spectrum of players concerned about the environment, including government administrators, farmers and ranchers, business leaders, philanthropists, social investors, and anyone who cares about bringing about a more prosperous future.

We humans can do astounding things when we focus on challenging problems. Not all the obstacles may be known or the details written down, but we have the tools we need to start the journey. What is certain is that we cannot afford to stay stuck in an extraction-based past that treats our natural resources as limitless. We must forge ahead and create a conservation-based future that balances a prosperous economy with a thriving environment. And we can. By reading this book, I hope you will walk away with a strong set of organizing principles with which to evaluate our present crisis and build a more resilient future.

CHAPTER 1

A New Conservation
for a New Era

Imagine walking into a job interview at a major manufacturing company. You've already gotten the tour of the administrative offices and are surprised by the absence of modern technology, let alone the large stacks of papers heaped on employees' desks. You've just completed the interview, answering all of the CEO's questions. Now it's your turn to ask some questions.

"Of all the widgets you manufacture," you begin, "which have been the most and least profitable?"

"I'm not sure," the CEO says.

You try to hold back your amazement. "Who are your biggest competitors?" you ask.

"Oh, there's a handful," she says, her voice trailing off.

You shift in your seat, trying to hide your discomfort. "What are your long-term goals for the company, and what threats could undermine your success?"

"We've been meaning to develop a business plan. It's just that we've been so busy managing our day-to-day affairs."

In today's world, it's hard to picture a business of any kind making these mistakes. What twenty-first-century corporation could survive if it neglected to define its objectives or analyze its progress? How long would it take for a business to tank if it failed to gauge market trends?

Yet this is exactly the way we approach our environmental problems. We lack real awareness of the situation; we don't fully understand the current state of our natural areas or what our actions might mean for their future. We don't precisely define our goals for improving the environment or use innovation and technology to help us achieve them. Nor do we adequately analyze our progress to make sure we're obtaining quantifiable results. It's like driving without a dashboard. We don't know how fast we're moving or whether we'll ever reach our destination.

The consequences of the current approach are devastating. Despite well-intended efforts by numerous environmentalists, policymakers, and philanthropists, the health of planet Earth continues to deteriorate at a startling rate. Sure, the environmental movement has won many notable battles. Yet, over time, the significance of these wins has declined to the point where we are now rapidly losing the war. Although today's environmental realities have changed, modern environmentalism keeps plugging away with the same outdated toolkit, and it is reaping an ever smaller return on its investment.

So how has the situation changed since the dawn of the environmental movement? Consider the following:

- World population has doubled to 7.2 million.
- U.S. population has grown by more than 55 percent to 316 million.
- The amount of pesticides used in the United States has tripled to 1.1 billion pounds per year.[1]
- The number of worldwide dead zones has spiraled from roughly 30 to more than 500.[2]
- The total number of freshwater species has declined by 50 percent.[3]
- Global Atlantic salmon catches have fallen by 80 percent.[4]
- Total acreage of U.S. wetlands has decreased by more than one third.[5]

- The amount of U.S. land consumed by urban development has doubled.[6]
- Annual carbon dioxide emissions that contribute to global warming have risen by more than 80 percent.[7]

If a time machine landed a human being from 1970 on today's planet, this passenger would find himself thrown into an almost unrecognizable world. The planet we live on today is dramatically different from that of a generation ago, when the modern environmental movement was born. To address today's realities, we need a radically different approach, not just an extension of the one we've used in the past.

To put it bluntly, we need to wake up and smell the future—because it's already here. We have entered a new environmental era, one with far more daunting problems than we faced 50 years ago. Yet we continue to muddle along like the manufacturing company described at the beginning of this chapter, doing things the same old way, failing to adapt to the new reality before us.

Adapting to the new reality requires implementing bold, innovative approaches that are a true match for the severity of the problems we face. It also means being adamant about obtaining results. For the environmental movement to continue to be relevant, it needs to remake itself into a more agile force that continually reevaluates the current situation and then adapts its practices to achieve the highest possible return on its conservation efforts.

It's not just environmentalists who need to change. Governments and philanthropists working on these issues need to get serious about demanding results. And agriculturalists and businesses need to recognize that it's in their own interest to conserve natural resources on which their livelihoods depend. If we're to survive a future in which 10 billion humans call planet Earth home, we must all work to solve our problems, and we need to begin now.

The good news is that, with the right focus and tools, we can achieve a more resilient environment. Think about all the human and financial capital that has been poured into the economy since the end of World War II. The resulting economic growth has been astounding. From 1950 to 2011, the gross world product—the combined gross national product of all countries in the world—has mushroomed from $7 trillion to $77 trillion.[8] These staggering numbers have come about because we've made financial gain our priority and consequently have spent the past several decades perfecting a set of practices that ensure businesses achieve the highest results.

Quantified conservation is about applying that same laser focus to achieve similar gains for the environment. It's about leveraging the best practices used by today's successful businesses and social sector organizations to overhaul the state of our natural resources. And it's about embracing the same sophisticated set of tools to bring about measurable improvements that ensure both a healthy environment and a thriving economy for decades to come. Simply put, quantified conservation is a twenty-first-century approach to solving the twenty-first-century problems that confront us. It offers a framework built on the following five principles, all of which the business world relies on for its success:

- *Situational awareness* to provide an objective understanding of the real-time environmental problems we face
- Bold *outcomes* that define the results we seek
- *Innovation and technology* to achieve our desired outcomes at the pace and scale needed for success
- *Data and analytics* to prioritize those environmental projects that have the most impact, measure our results, and monitor our progress
- *Gain*, which becomes the threshold question for public, private, and philanthropic investment by tying every dollar to measurable net benefits achieved for the environment

Although the chapters that follow are about using quantified conservation to improve the future of water, these same principles can be used to tackle any environmental problem. Whether the goal is saving our forests, restoring the diversity of wildlife, or reversing the effects of climate change, quantified conservation can address our environmental problems with far greater precision and sophistication. It's that kind of focus that will be needed if we're going to maintain a healthy environment in our twenty-first-century world—one in which we're bumping up against the limits of our natural resources with greater frequency and severity.

Situational Awareness

There's an old adage that says you never step into the same river twice. Yesterday's water that flowed past the point at which you're standing has long since moved on. In the same way, our world is always changing. The pace of change in the twenty-first century is incredibly fast and continues to accelerate. To get an accurate pulse of the current situation, we need to continually monitor it so we can constantly reevaluate where we stand and quickly make the necessary changes.

Companies that don't do this well end up losing market share or eventually go bankrupt. Consider the Canadian smartphone maker BlackBerry, a company that revolutionized mobile devices. Just a few years ago, BlackBerries were so popular that users called them "CrackBerries" because of their addictiveness. Yet BlackBerry failed to keep pace with market trends. Having built its success on keyboard-equipped mobile devices, the company failed to anticipate the consumer desire for touchscreens. It lacked the agility to stay ahead of the competition, and in 2013 it was forced to sell because of its declining financial position.

Now consider a company like Apple, which went from near bankruptcy to billions within a decade.[9] With a focus on out-of-the box concepts ranging from the iPod to the iPhone to the iPad, Apple anticipated trends in the rapidly changing market and introduced a chain of must-have products that mirrored its "think different" motto. The key was consistently adapting itself. Rather than resting on its laurels, Apple continued to gauge the market and create new innovations that customers wanted.

As Apple's Steve Jobs put it, "If you do something and it turns out pretty good, then you should go do something else wonderful, not dwell on it for too long. Just figure out what's next."

The unfortunate reality is that modern environmentalism has gone the way of the BlackBerry. With its focus on advocacy and litigation, the environmental movement initially hit some huge home runs. Yet our streams and rivers are now in many ways worse off than when the Clean Water Act was passed nearly a half century ago.

Today, more than half of the 3.7 million miles of streams in the United States are polluted or damaged. Take the Colorado River, for example. This once mighty river system that provides water to 40 million Americans has lost so much water that it now rarely reaches the ocean. Likewise, the iconic Mississippi River has become so polluted that it's created a massive "dead zone" in the Gulf of Mexico that's literally choking out fish and other marine life. Unfortunately, dysfunctional rivers such as the Colorado and the Mississippi are now the rule in America, not the exception. Simply put, water is in trouble. And when water is in trouble, everything is in trouble, including the economy, the environment, and life itself.

With so many environmental groups and policymakers working to protect U.S. rivers, how can our watersheds be headed for crisis? It's not that Conservation 1.0 hasn't had its successes. After all, it brought about the U.S. Clean Water Act, which has virtually eliminated point source pollution—or pollution discharged into rivers directly from factory and sewage plant pipes. Thanks to these early advocacy efforts, we no longer have fires burning on rivers, as on Ohio's Cuyahoga River. That was a huge victory, and today's rivers are certainly better off for it.

Yet, in the meantime, a new set of problems have emerged. The major water pollution problem we now face is nonpoint source pollution—mainly fertilizers, herbicides, and insecticides from agricultural production that seep into the water from multiple points along rivers and streams. Add to that skyrocketing population and an advanced economy, both of which are putting unprecedented pressure on our limited water supplies. And add to *that* the worsening megadroughts and intense flooding created by climate change, which are making our water supplies less and less predictable.

For The Freshwater Trust, the realization that the return on our conservation effort was diminishing came in 2002, when I completed an analysis to better understand our successes and failures. We had begun as a fish advocacy organization, and we had made some important wins along the way. We had successfully listed several of the first Pacific salmon under the Endangered Species Act. We pioneered the first water trust in the nation, using the model of the land trust to protect streams and rivers. And we had established a reputation as a small but effective conservation group.

Yet increasingly, it felt like we were fighting over commas in fish management policies. We were no longer moving the needle for wild fish. In fact, we were barely holding the line against fish declines.

We saw the imperative to redirect our focus from fish advocacy to river restoration because that's where the metrics showed we could achieve the greatest gains for the environment. As it turns out, the decision was spot on. As we continue to analyze our progress, it's clear that we are having far more impact by adopting this new focus, with the potential for even greater impact in the future.

Situational analysis starts at the organizational level. It's essential that environmentalists, policymakers, and philanthropists—and anyone who plays a role in protecting our watersheds—continually step back to assess the current situation and evaluate the return they're getting on their conservation investment.

At the project level, too, we must understand the changing situation of specific river systems in order to fully understand what must be done to fix them. Soaring population, climate change, changing land use, and economic advances are altering our rivers and streams all the time. How well and how quickly we react matters. Being situationally aware requires continually monitoring the state of our streams and rivers to obtain an objective understanding of the present reality. Situational awareness at both the organizational and project levels is a critical first step if we are to address the magnitude of the freshwater issues we face.

Outcomes

Once we understand the state of our watersheds, we can then establish concrete goals. Given the current rate of deterioration, we cannot tinker our way to success. The outcomes must be clear and ambitious—and at first blush, they should sound unreasonable. As escalating population puts increased demand on our rivers and climate change creates unpredictable water supplies, incremental improvements won't be enough.

In setting outcomes, we can learn a lot from Google's Larry Page, who believes that incremental improvements of 10 percent don't do much beyond furthering the status quo. Instead, what Page expects of his employees is that they create products and services that are 10 times better than the competition. As he told *Wired* magazine, "Thousand-percent improvement requires rethinking problems entirely, exploring the edges of what's technically possible, and having a lot more fun in the process."[10]

In the same way, we need to set our sights high. We need to establish bold outcomes that demand thousand-percent improvements over the status quo. In short, we need to live by the gospel of 10x. This may seem like a tall order for the environment, but it's standard practice among the world's most successful companies.

Take Amazon, for example. In 2004, the company set a goal for its series of Kindle e-book readers to make available "every book, ever printed, in any language, all available in less than 60 seconds."[11] Setting a clear and ambitious outcome enabled Amazon to establish a well-defined focus, and, not surprisingly, the results have followed. Today there are more than 1 million books, magazines, newspapers, and blogs available on the Kindle under 60 seconds. Kindle sales continue to surge and are expected to reach a whopping $5.5 billion by 2015.[12]

So what outcome has traditional environmentalism set for itself? For the most part, it's been to stop things from getting worse. The vast majority of money and effort isn't being applied to undoing the harm already created or making positive gains happen but to stop further environmental damage from happening. Imagine Amazon or any other business establishing for itself a goal this timid. No improvement, no innovation—just holding the line. It wouldn't take long before a company would be trounced by its competitors.

As can be seen from all the black lines in figure 1.1, damaged rivers abound in the United States. With more than half the miles of our streams and rivers already in trouble, simply holding the line amounts to a weak outcome. To make real gains for the environment, we need to change those black lines to a healthy blue. We need to take rivers that have already been damaged and restore them to their natural state so that we have enough clean freshwater for both a sustainable economy and a sustainable environment.

On a large scale, the bold outcome would be to turn all black lines to blue within a decade. To do that, a prerequisite would be to understand by watershed the specific quantity and quality of water needed to sustain a healthy ecological system while meeting the needs of agriculture, industry,

Figure 1.1. Areas across the United States with impaired rivers and streams. *(Credit: The Freshwater Trust, using data from U.S. Environmental Protection Agency, 303(d) Listed Impaired Waters. NHDPlus Indexed Dataset with Program Attributes, August 4, 2014.)*

and urban users. All this information and more should be available in real time. Yet we're nowhere near that. In fact, it took dozens of government agencies the better part of a decade to compile this map—unacceptable in a world where we can get almost any book ever written, in any language, in our hands in less than 60 seconds. As businesses such as Amazon realize, outcomes can be powerful because they provide a focus and set the design parameters.

Innovation

Equipped with bold outcomes, we can turn our attention and our talent to achieving them in the most effective way—which isn't always the traditionally accepted way. It's not enough to simply work harder. We must work smarter, and innovation can help us do that by offering ways to get things done more effectively. Innovation can take many forms. It can be a new method or way of doing things, a new technology or tool, or a new form of interaction. It can also be a previous advance adapted to fit a new need.

In the business world, innovation is a relentless imperative. Businesses understand that they must continually innovate to hold onto their competitive advantage. As Jack Welch, the former CEO of General Electric, once said, "I take no comfort in where we are today." Business leaders understand that they must continue to innovate in order to survive.

Rated by *Fast Company* magazine as one of the world's most innovative businesses, Amazon is always searching for ways to boost its efficiency. The company realized it could double its storage space by stocking its warehouses in a dramatically different way.[13] Rather than grouping the same items together, Amazon now uses the "chaotic storage" system in which items are barcoded, tracked in a database, and shelved wherever there's empty shelf space.[14] The new method is far more effective, yet it was far from intuitive. It was achieved only by proactively rethinking the company's day-to-day operations.

Amazon also plans to operate more efficiently by adapting existing technology. To provide same-day delivery service to its customers, the company is researching the use of drones to deliver its products. The flying machines, which Amazon calls "octocopters," can carry packages within a 10-mile radius of each Amazon warehouse, making deliveries within just 30 minutes from the time they were ordered.[15] Although Amazon's innovations may seem futuristic and perhaps even bizarre, they illustrate the out-of-the-box thinking, focused on the end result, that's needed to help us solve some of the toughest problems facing the environment.

As with Amazon's octocopters, innovation often comes from adapting existing technology to fit a new need. In other cases, innovation can be something that seems altogether impossible. For example, when President Kennedy set the goal of "landing a man on the moon and returning him safely to the earth" by the end of the decade, he didn't know exactly how we'd get there. He was simply driven by the fact that the Soviets had a head start on the space program, and he wanted the United States to get there first. By setting an ambitious outcome, Kennedy created a clear focus for the country, and the technology that made it all possible followed as a result.

In the same way, innovation will naturally follow as environmentalists, governments, philanthropists, businesses, and others who influence our natural resources set stronger environmental outcomes. The opportunities for innovation are boundless. From water-saving irrigation technologies to no-waste design certifications such as "cradle to cradle," we are already starting to see innovations that conserve our natural resources. As we bump up ever harder against the limits of freshwater, innovation will play an increasingly vital part of the solution.

Data and Analytics

Despite their importance, situational awareness, outcomes, and innovations aren't by themselves enough. Success requires data and analysis. Today, there's an incredible amount of data floating around the web and other digital media. In fact, every day we create 2.5 quintillion new bytes of data.[16] That's so much data that if each of those bytes were pennies, we'd have enough pennies to completely cover the surface of the earth five times over every single day.[17]

Although the tools are there, we have just started to scratch the surface for how data and analysis can be used to improve our watersheds. The use of data and analytics in the environmental world is like the invention of the chronometer for calculating longitude on a ship. The chronometer was developed in the early eighteenth century, for the first time making it possible for navigators to calculate the exact position of their ships by accurately measuring time and distance. Yet despite their ability to prevent ships from becoming lost at sea, these instruments weren't put into widespread use for another 40 years because of politics and an unwillingness to change.[18] That's 4 more decades of losing lives at sea when the technology already existed to prevent such tragedies!

In the same way, we have the tools needed to steer our watersheds back to health. We just need to put them to use. Addressing the sheer scale of freshwater issues we face demands that we move toward a quantifiable, data-driven existence.

In the business world, data and analysis have been steering the ship for decades. According to one survey, 73 percent of companies leverage data to increase revenue, and 84 percent of executives use data to help them make better business decisions.[19] On a large scale, the financial world embraces economic indicators to measure the health of the overall economy. On a smaller scale, businesses use data and analysis to identify new opportunities, prioritize their efforts, refine their methods, and measure their progress toward their goals.

In the same way, data and analysis can be made to work for the environment. One organization that's harnessing data and analysis to better manage water is the Sonoma County Water Agency. Located in the heart of northern California's wine country, the public agency needed a way to conserve water as population growth and increased drought were straining its resources. To do that, it teamed up with IBM to develop a sophisticated water management system based on near real-time information. The system includes dashboards and maps that help Sonoma County quickly identify potentially defective water pipes so that it can repair them before problems occur.[20] The system also provides the agency with up-to-date information about water usage and quality, weather and climate, and environmental considerations so that it can allocate water more effectively. Using data and analysis to inform its decisions, Sonoma County is reducing water usage at a time when California's water supplies are dwindling.[21]

On a macro level, too, we need to make much better use of data and analysis. Just as we rely on economic indicators such as the gross domestic product (GDP) and the consumer price index (CPI) to measure the health of the economy, we need a set of indicators to assess the health of the environment. Some environmental health indicators already exist. For example, Yale University and Columbia University issue an annual environmental health index (EHI) that ranks how countries perform in two areas: protection of human health from environmental harm and protection of ecosystems.[22] In addition, the Organisation for Economic Co-operation and Development (OECD) teams up with other international organizations to publish environmental indicators in areas such as water consumption, fisheries, and carbon dioxide emissions.[23] Just imagine if indicators like these were taken as seriously as the Dow Industrial Average!

Indicators are also needed at a more granular level. Just as businesses use budgets and accounting to track their spending, indicators can help us define the limits to a healthy river system. What is the total maximum daily load (TMDL) of nutrients that a particular river system can absorb and still be fishable, swimmable, and drinkable? What's the total amount of water needed for a thriving river basin? What is the total allowable catch (TAC) that preserves a healthy fish population for tomorrow?

Using data and analysis, we can now measure these limits, representing a revolutionary shift in how we manage natural resources. For the first time in human history, we can understand the environmental boundaries within which we can safely operate. We have a set budget that tells us how much we can use and the accounting tools we need to remain within budget. Understanding how much a river can handle, we can make the informed tradeoffs needed for both a functioning economy and a functioning environment.

As we go about improving rivers, we can also use data and analysis to inform our restoration work. In the past, we've measured our restoration efforts by the physical tasks that were performed—the number of trees that were planted, for example, or the acres of wetlands preserved. In recent years, however, scientists have begun to quantify the benefits nature provides as "ecosystem services." For example, an acre of native shrubs can absorb X tons of excess nitrogen rather than send it into the stream. A tree can sequester Y amount of carbon. And an acre of wetland can cool water by Z amount of heat. Measuring ecosystem services radically improves restoration efforts by showing us exactly what must be done to achieve the outcomes we've established.

Likewise, data and analysis can help us pinpoint where exactly to focus our efforts. For example, The Freshwater Trust prioritizes its tree planting restoration projects using Basin Scout, a tool that examines vegetation, water temperatures, fish populations, and other factors along a river, to identify which specific areas have the most to gain from shade. By using the Basin Scout, we've been able to focus our efforts on the most important restoration projects, helping us achieve the greatest results at the least cost.

Finally, data and analysis can also be used to monitor the success of our river restoration projects after their completion. With sophisticated tools, we can gauge whether our efforts are achieving the desired impact over time. Are the trees growing as planned and providing the watershed cooling benefits as modeled? How must we adjust planting designs to maximize the environmental benefit while optimizing costs? Analytics can take the guesswork out by absorbing and synthesizing astounding amounts of information. For example, The Freshwater Trust developed a software application that enables workers to enter data right from the field as they go about monitoring projects. The data are entered via an iPad and then transferred to a database at our headquarters, where the numbers are crunched. With this monitoring app, analyses that used to take days now take minutes, allowing us to quickly make adjustments to river restoration projects based on near real-time information.

Interestingly, data and analysis tell us whether we are making the needed environmental gains and feed back into situational awareness. In fact, the five principles of quantified conservation form a continuous loop (figure 1.2). With consistent information about the state of our watersheds, we have real-time information about the situation we face, which allows us to update our outcomes and embrace new innovations to tackle the latest realities.

Gain

Traditional investment principles must expand into environmental effort. To make up for lost ground, we need massive amounts of restoration. And the immutable truth is that restoration takes cash. We've got to pay for

Figure 1.2. Quantified conservation is a continuous process. *(Credit: The Freshwater Trust.)*

these projects with both public and private dollars. Quantified conservation allows us to stretch existing dollars by tying them to measurable gains achieved for the environment. It also paves the way for market mechanics to channel additional funding into restoration while using proper design to balance the needs of the environment with those of the economy. By attaching every dollar spent to measurable benefits for ecosystems, we enable the transformation of environmental spending.

Public and Private Spending

It's hard to imagine any successful twenty-first-century business spending money on a program without measuring its results. Yet when it comes to environmental funding, that's exactly what we're doing. Each year, we spend billions of public and private dollars on freshwater ecosystems in the United States without truly knowing what we're getting. Often, we don't have a solid understanding of the situation we're trying to fix, we don't define our outcomes or use technology to help us reach them, and we don't use data and analysis to measure our success.

At the federal level, for example, the U.S. Department of Agriculture budgets roughly $6 billion a year for programs that enhance water resources while encouraging the conservation and restoration of private land and national forests.[24] Other habitat restoration programs totaling hundreds of millions of dollars are funded by a hodgepodge of federal agencies, including the U.S. Environmental Protection Agency, U.S. Department of Commerce, U.S. Department of the Interior, U.S. Department of Health and Human Services, U.S. Department of Defense, and U.S. Department of Transportation. These dollars come in addition to millions of dollars in private grants made to nonprofits to protect our watersheds. Just think if all that money were spent on measurable improvements to the environment. Our rivers and streams would be in much better shape than they are today.

Instead, what usually takes place is a very subjective process. Whether seeking a grant from a government agency or a private philanthropist, environmental groups typically submit a grant proposal that includes a description and cost of the project and an explanation for why it's needed. The grant maker evaluates the proposal alongside many others, comparing the cost and merits of the project based on description alone. Without quantified conservation, grant makers have no way of knowing whether they've chosen the best project. Nor do they know whether the project achieved its intended result. They're simply informed when the project is completed, sometimes along with a report describing the great work of the organization.

Quantified conservation has the ability to revolutionize philanthropy and public investment by allowing funders to be far smarter with their investments. No longer will grant makers fund conservation. Instead, they'll purchase quantified outcomes. Rather than paying for the planting of 5,000 trees, for example, grant makers will be able to buy a 10-million-kilocalorie reduction in watershed cooling shade—and they'll make this purchase only after the work has been completed and verified by a third-party auditor. What's more, they'll be able to monitor the effect of the trees on watersheds over time. With a quantified way to measure the benefits of restoration projects, grant makers will have an objective way to measure the success of their investments.

In addition, quantified conservation will help funders prioritize their investments. For example, if a grant maker receives two proposals, both of which involve planting 5,000 trees at the same price on the same river, it would appear that these were equally valid projects. Yet with quantified conservation, grant makers will be able to compare the two projects on a far more granular level, taking into consideration factors such as the slope of the bank, the angle of the sun throughout the year, and the length of the river channel that will be covered by shade—in order to get the greatest amount of environmental benefit for their dollar. Using the right tools, we can measure the specific benefits of both projects over time. Perhaps one project lowers heat from the sun by 10 million kilocalories per day, compared with 5 million kilocalories for the other project. In addition, using digital maps to chart water temperature and fish habitat along the river, we can conclude that the 10-million-kilocalorie reduction project would add shade at a location where it's needed more.

Quantified conservation provides a powerful new way for investors to target their limited funds to the projects with the most impact. By quantifying the environmental improvements of every public and philanthropic dollar, we will be able to accomplish far more for our rivers using the same amount of money.

Environmental Markets

In addition to making existing conservation dollars stretch further, quantified conservation can attract additional money into river restoration by providing the tools needed to design successful environmental markets. In simple terms, a market means trading. And in the case of environmental markets, a negative environmental impact is traded for an environmental improvement. By properly quantifying environmental impacts and improvements and turning them into measurable units, quantified conservation paves the way for informed trading to take place.

Environmental markets aren't new. In fact, they've existed for decades. From markets that reduce acid rain to those that protect wetlands, environmental markets can serve as an effective conservation tool that balances the needs of the environment with that of the economy. Yet to be successful, these markets need to be designed in a way that brings about measurable, lasting gains for the environment rather than transactions that simply monetize nature and turn a buck. With better quantification tools now available, environmental markets are poised to do just that.

In the long run, environmental markets will bring additional money into river restoration by turning conservation into a sound investment opportunity. Measurable units of environmental gain will be traded for profit, and the need to restore rivers will spur a whole new class of nonprofits that perform on-the-ground work with money left over for additional projects that benefit the environment.

In the short run, environmental markets will encourage industries regulated by the U.S. Clean Water Act to meet their environmental obligations using natural infrastructure rather than built architecture. Historically, industrial permit holders have cooled their clean water before returning it to the river by purchasing cement cooling towers, refrigerated chillers, and other costly solutions that comply with the rules but do little to improve the environment. By buying units of environmental improvement that result from efforts such as planting trees and restoring wetlands, they can instead channel this money into projects that actually restore rivers while saving money.

In addition to water temperature, water trading programs can be designed to reduce other pollutants, such as the amount of phosphorus, nitrogen, or sediments that seep into watersheds from agricultural production. They can also be developed to manage limited water supplies in a way that encourages conservation and channels water toward its most important uses while ensuring that enough water remains in-stream to support fish and other river habitat.

The time is ripe for investments that create gains for the environment, and not just the stock market. If we can steer public funding, philanthropic dollars, and environmental market investments toward efforts that have the greatest measurable impact, we can make transformational gains for our natural ecosystems in a way that traditional methods no longer can.

Every Generation Needs a Revolution

Thomas Jefferson once said, "Every generation needs a new revolution." Quantified conservation offers a revolutionary approach to managing our water and all of our natural areas—one that's sorely needed if we are to maintain sufficient resources to meet our current and future needs. With *situational awareness*, we can obtain an accurate picture of today's ecosystem realities and from there set *outcomes* that move the needle in favor of the environment. Through *innovation*, we can pursue the results we need to hit our needed outcomes. With *data and analytics*, we can precisely inform our environmental work, better prioritize our efforts, and measure our progress toward our goals. And by committing to *gain*, we can spur the investment needed to achieve our outcomes at a pace and scale needed for success. By harnessing the everyday methods used by today's most successful companies to obtain measurable gains for the environment, quantified conservation can meet the needs of the economy while improving the resilience of our planet for decades to come.

By adopting the principles of quantified conservation, all the key players who influence the environment—including governments, environmentalists, agricultural interests, businesses, and investors—can bring about a future that is both economically and environmentally prosperous. Yet to do that, we need to shift our thinking, our actions, our capitalism, and our environmental effort to an altogether new mode. As we will see in the next chapter, the first step is ending our open-ended accounting system—and that requires us, once and for all, to face up to the reality that freshwater and all of the planet's ecosystem services are indeed limited.

CHAPTER 2

Leading in a World of Permanent Scarcity

KLAMATH RIVER BASIN is ground zero when it comes to the battle over water rights in the United States. Flowing more than 250 miles southwest across Oregon into northern California, the Klamath once supported the third largest salmon run in the nation.[1] Yet the construction of several dams has weakened water flows as water is diverted for irrigation purposes. As drought and overuse have further shrunk available water, farmers and ranchers for more than a decade have squared off against commercial fishers and Native American tribes for the limited supplies needed to support their livelihood. The battle has been brutal on all parties and at times has bordered on violence.

In the meantime, both the river and the local economy have suffered. Water shutoffs have threatened the cattle industry while leaving farmland fallow. River flows have been so low they've caused massive fish die-offs. Recurring toxic algae blooms have fouled reservoirs, and salmon population declines have closed 700 miles of coastline to fishing. At one point, former vice president Dick Cheney ordered the Interior Department to provide water to farmers regardless of river conditions, causing tens of thousands of salmon to wash up on the Klamath River's shores. It was the largest salmon die-off in the entire history of the Pacific Northwest.[2]

On the other side of the United States, another crisis has been brewing, this one affecting the Chesapeake Bay, the largest estuary in the nation and the third largest in the world.[3] The Chesapeake, which covers 64,000 square miles and touches six states (Maryland, Virginia, Delaware, West Virginia, Pennsylvania, and New York) and the District of Columbia, was once one of the most beautiful and productive estuaries on Earth. It was home to more than 3,600 species of animal and plant life and provided important resources to the local economy, especially crabs, oysters, and rockfish.[4] Yet today it is one of the top waterways on the U.S. Environmental Protection Agency's "dirty water" list.[5]

Over time, the activities of 17 million people have overwhelmed the bay. The Chesapeake has been stripped of forested buffers and wetlands, allowing the growing quantity of chemicals and other pollutants to seep unfettered into the bay. The largest source of water pollution has been nitrogen from agricultural runoff, specifically chemical fertilizers, livestock manure, and sewage sludge on fields as well as from animal waste that spills off of pastures and feedlots.

The result has been a growing number of dead zones, or areas of little or no oxygen where fish, crabs, oysters, and other marine animals suffocate. Today, fish and shellfish populations are a fraction of what they once were.[6] The decline has taken a toll on the $3.3 billion local commercial fish industry, which supplies 34,000 jobs to the local economy.[7] The ecological disaster has also hampered recreational fishing, which contributes more than $1.6 billion to the region's livelihood.

Victims of Our Outdated Water Laws

The Klamath River Basin and the Chesapeake Bay, at opposite ends of the United States, are representative of the types of water shocks that are erupting all across the country. Such crises aren't just isolated instances. Long in the making, they are popping up in nearly every region of the country, with devastating consequences for both our economy and our environment. These ecological disasters are not natural functions gone awry but rather the manifestation of ineffective U.S. water management policies, some of which have been in place for centuries.

The Klamath River Basin is governed largely by the "prior appropriation doctrine," a 150-year-old policy that took root when European Americans first settled the West and the land was sparsely populated. Miners searching for gold treated squabbles over water as they would disputes over minerals: "First in time, first in right," meaning that the first miner to use the water had a right to continue using it, to the exclusion of others. At its core, it was a practical way to define who got there first and "captured" the resource for their own use. In new territories, where municipal clerks were often days away and no databases were handy to search land ownership, a sign could be tacked up to a tree notifying others that this stream was already under claim—and often the late arrivals would simply move on.

Similarly, water pollution in the Chesapeake Bay is partly the consequence of our outdated 1972 U.S. Clean Water Act. Although the Clean Water Act protects rivers and streams from point source pollution, or industrial pollution from pipes, it does little to regulate nonpoint source pollution, or pollution from diffuse sources, such as agricultural runoff, which in the 1970s was a fraction of the problem that it is today. Today, the use of agricultural pesticides and fertilizers has skyrocketed, and nonpoint source pollution from agricultural runoff has emerged as the number one water quality problem facing U.S. watersheds. Yet despite several updates, the Clean Water Act has never been able to address this problem.

Like the Klamath River Basin and the Chesapeake Bay, watersheds across the United States are increasingly the victims of outdated and fragmented water laws that have failed to keep pace with twenty-first-century realities. Many of our laws were developed at a time when freshwater was just another factor in our pursuit of prosperity. And they were instituted before industrial agriculture, fertilizers, and pesticides came into widespread use.

Open-Ended Accounting on a Finite Planet

In 1968, astronauts on the first human mission to the moon captured a photo of a tiny Earth enveloped by vast space (figure 2.1). Declared "the most influential environmental photograph ever taken," the photo of Earth captured the imagination of people around the world, giving them a new perspective on the fragility of our planet. It was perhaps the best visual argument that Earth is indeed finite and its natural resources limited.

If we were ignorant of these facts before the picture, we've largely disregarded them over the last half century since the photo was taken. Population has almost doubled since 1970, and with it the amount of Earth

Figure 2.1. The Apollo 8 spacecraft captures the Earth rising above the lunar horizon. The first photos taken of Earth from space present a visual argument that we live in a closed-loop biosphere with finite natural resources. *(Credit: NASA.)*

per person has shrunk significantly. From information derived from the United Nations Human Development Index and Ecological Footprint, Malcolm Preston, global head of PricewaterhouseCoopers's sustainability practice, incisively makes sense of what's happening on our planet (figure 2.2).

The graph shows that not a single nation is above what is effectively the United Nations poverty line *and* living within the sustainable limits of the earth. The country's either poor and below the earth's carrying capacity or wealthy and living beyond what the earth can handle. When you look at the graph, you think to yourself, "We're screwed." And if all we do is more of the same, we certainly are screwed. But Preston sees more. He sees the greatest economic opportunity in the history of humankind—as long as we properly account for inputs, outputs, and leftovers. He is an accountant type who has spent a lifetime doing deals, after all.

But Preston is right: Every challenge is an opportunity. And governments at all levels across the United States could be leading the way—if they only acknowledged that the earth has a budget and that we must

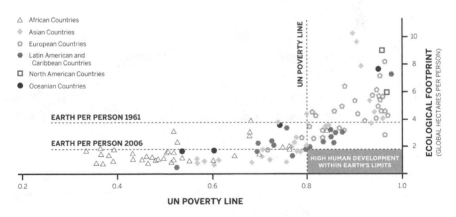

Figure 2.2. Countries outside the lower right hand box are unsustainable, either environmentally or economically. *(Credit: The Freshwater Trust, using data adapted from U.N. Human Development Index and Ecological Footprint, 2006, as published in "The Ecological Wealth of Nations. Earth's Biocapacity as a New Framework for International Cooperation," Global Footprint Network, April 2010, figure 7, p. 13.)*

live within it. One of the reasons no nation is where it should be is that policymakers around the world continue to govern based on the false assumption that economic growth can continue at the expense of the environment and do so ad infinitum. As Richard Heinberg put it in his excellent book *The End of Growth*, this endless pursuit of growth that ignores environmental limits amounts to "a flight from reality."[8]

Yet the tunnel vision continues, and the myriad environmental laws that we have on the books reflect this outdated thinking. In some cases, our water laws suggest a lack of planning. In other cases, they reveal a hodgepodge of policies fashioned over multiple eras. In almost every case, they trail far behind today's realities by failing to demand a full accounting of our actions on a finite planet.

Piling Up a Heritage of Conflict

Geologist and explorer John Wesley Powell is considered one of America's earliest climate scientists. In the late 1800s, when he was sent to survey the West for the federal government, he observed that areas west of the 100th meridian—a north–south demarcation running from the Dakotas to Texas—are simply far more arid than the East. Convinced that the lack of rain would be an ongoing problem in America's westward expansion, Powell proposed that political boundaries be organized around watersheds and managed as commonwealths (figure 2.3). By forming communities around rivers, he reasoned, the early settlers would see firsthand how little water existed and therefore work to conserve it. What's more, communities would be joined by the common goal of managing a scarce resource that would be shared by all.

Yet the U.S. Congress saw Powell's plan to end the free-for-all distribution of land as too much surveying, planning, and regulation.[9] America was in a hurry at the time, and many leaders thought it would interfere with the rapid development of the West. In the end, congressional leaders gave in to rail companies and other business interests that were banking on large-scale settlement and agricultural development. As one supporter of rapid development put it at the time, "The rain follows the plough."[10]

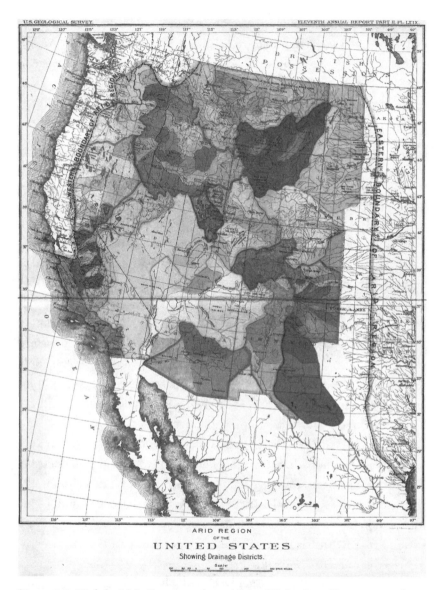

Figure 2.3. Had the U.S. Congress adopted John Wesley Powell's recommendations, the West would have been set up to better manage its water supplies. *(Credit: John Wesley Powell's 1890 map of the "Arid Region of the United States: Showing Drainage Districts," published in the U.S. Geological Survey's 11th Annual Report, 1890–91.)*

Knowing that simply plowing the desert floor would not coax rain, it's easy to see why Powell, a man of scientific bent, was frustrated by this wishful thinking. As he presciently warned at an irrigation conference in 1883, "Gentlemen, you are piling up a heritage of conflict and litigation over water rights, for there is not sufficient water to supply the land."[11]

Putting the Incentives in the Wrong Place

As we now know, states west of the 100th meridian were carved up according to straight lines on a grid rather than following the natural contours of the land. This largely divorced the water from the land across the region and ensured a disintegrated management of the resource. Moreover, Powell was right in his prediction that water would become a huge source of conflict between western states. Consider the conflicts that have erupted between western states surrounding Lake Powell (named after John Wesley Powell, ironically), whose water levels have dropped to record lows. Sadly, water in the West wasn't allocated in the planned manner that Powell had envisioned but rather as a free-for-all in which the first to claim it got it, regardless of whether they lived next to the river. The only restriction to the prior appropriation doctrine was that water rights holders put the water to "beneficial use," which was broadly defined as having some economic, industrial, or household benefit.

East of the 100th meridian, where water was more plentiful, water use followed a different path. Here, most states adopted the riparian doctrine, which was modeled on English common law and prohibited altering the course of water except for basic consumptive uses. The riparian doctrine took a more communal approach to water allocation, holding that every riparian landowner has the right to have water flow past their land undiminished in quantity or quality.

In both the eastern and western United States, most states still follow some version of the prior appropriation doctrine or riparian doctrine. And in most states, these laws are sorely outdated and unevenly applied. Although water users in riparian doctrine states are now required to obtain state water permits, doing so typically isn't difficult, even in areas where water levels are decreasing or unknown. What's more, most states are exempt from regulation water uses under a certain threshold, typically 100,000 gallons per day.[12] In the meantime, water rights holders in the western United States are effectively encouraged to waste water, since failing to use all of one's water can mean forfeiting the water right. Even worse, the total amount of water that's been granted on paper in the form of water rights *far exceeds* the amount of water that actually exists in our watersheds. In other words, on paper we're water rich, but in stream we're often water poor. It's akin to believing that you have money in the bank simply because you have checks left in your checkbook.

Although many states have made small modifications to the prior appropriation and riparian doctrines, most fail to require a quantified approach. For example, most states don't mandate that we monitor how much total water we have to work with. Nor do they require us to measure how exactly that water is being used. A century and a half after these water doctrines were put into place, we still don't monitor or measure, and as a result, we continue to operate in the dark.

Our Groundwater Free-for-All

Our laws governing surface water encourage users to waste water, but groundwater essentially isn't regulated at all. As with surface water laws, most states' groundwater laws evolved at a time when unrestricted pumping was widely viewed as acceptable and when little was known about the science of groundwater. *Frazier v. Brown*, an 1861 Ohio Supreme Court decision that rejected any legal constraints on groundwater pumping, illustrates our historical misunderstanding about groundwater. The case concluded that the existence, origin, movement, and course of groundwater are "so secret, occult, and concealed" that attempting to create any rules regulating it would be hopelessly uncertain.

Without interference from the courts, landowners pumped without restraint from nearby well users or public regulators. In 1900, nearly every state followed the "rule of capture," which was basically a free-for-all in which landowners could use as much groundwater as they liked, even monopolizing it, without liability.

Since then, science has advanced considerably, yet little has changed with respect to groundwater legislation. Today, groundwater hydrology is no longer the mystery it once was, and we can accurately predict groundwater recharge rates. Although some regulations have since been added, most states across the United States still allow unlimited access to groundwater.[13] And even the states that restrict groundwater pumping exempt domestic wells from regulation. In addition, actual well water consumption in most places isn't monitored or reported, so there's nothing in place to help governments quantify how much well water individual well owners are using.

With such weak policies in place, the result is that groundwater tables have been rapidly declining. Groundwater depletions in the United States doubled from 1950 to 1975 and have continued to escalate well into the current century.[14] If the current pattern continues, some parts of the country, such as the U.S. High Plains and California's Central Valley, won't be able to support irrigated agriculture within the next few decades.[15]

With one quarter of the nation[16] and 42 percent of farmers and ranchers[17] dependent on groundwater for their water supplies, the long-term results could be devastating. As Dave Owen, a University of Maine law professor, put it, "We generally give groundwater very little thought. Because it is concealed from view, most people have only vague, and often inaccurate, conceptions of what groundwater is, where it comes from, and how it moves. But obscurity does not mean unimportance. Groundwater plays a central role in our daily lives."[18]

A Tangled Pile of Water Quality Laws

Not only does the quantity of water play a central role in our daily lives, but so does the quality. Yet here also the lumbering pace of government has failed to keep up with our twenty-first-century realities. Remember the game pick-up sticks? A bunch of wooden sticks are dropped on the table, falling into a tangled pile. Each player then takes a turn picking up a single stick, trying not to move any of the others. In a sense, our water quality laws are like a tangled pile of pick-up sticks. They're piled up in random disarray. And it's nearly impossible to deal with one law without trying to contort your way around all the others.

At the federal level, water quality is touched upon, either directly or indirectly, by several landmark pieces of legislation. For example, the Clean Water Act regulates the discharge of pollutants into U.S. waters. The Endangered Species Act protects threatened and endangered species and the ecosystems on which they depend. And the Safe Drinking Water Act ensures the quality of America's drinking water. Although these are all good laws, good laws without proper integration add up to little more than uncoordinated chaos.

A case in point is the hodgepodge of federal agencies that share administration of water issues. For example, the U.S. Environmental Protection Agency has the primary authority over point source pollution. The U.S. Army Corps of Engineers is in charge of regulating wetlands and issuing permits for building dams. The U.S. Bureau of Reclamation manages large-scale irrigation systems in the American West. The U.S. Fish and Wildlife Service oversees the protection of endangered aquatic species, as does the National Marine Fisheries Service, at least for species that spend some time in the ocean. The U.S. Food and Drug Administration oversees drinking water safety standards. The U.S. Geological Survey collects data about the health of our watersheds. The U.S. Department of the Interior acts as a trustee for federal and tribal water rights. And these are just a few examples.

If that weren't complicated enough, federal funding for water is split across thirty agencies and programs, whose budgets vary widely along with changing administrative priorities.[19] The problem with this piecemeal approach is that few of these agencies' central missions revolve around water. As a result, no one agency is examining water issues in an integrated manner that takes all the different uses and potential conflicts into account.

Even the Best Laws Eventually Become Outdated

Unlike water allocation doctrines that date back centuries, the first strong federal water quality laws were implemented in the early 1970s, spurred in part by growing public support for clean water in the wake of Rachel Carson's landmark book *Silent Spring* and shocking photos in the national media showing fires burning on Ohio's Cuyahoga River (figure 2.4). Responding to the public outcry, Congress in 1972 passed the U.S. Clean Water Act to replace ineffective state regulation of pollution with a more comprehensive national system. The goal of the Clean Water Act was to "restore and maintain the chemical, physical, and biological integrity of the Nation's waters."

The Clean Water Act made it illegal to dump point source pollution (pollution from pipes) into the nation's waters unless users obtained a federal permit. The idea was to decrease these pollutants over time using the best available pollution control technologies, with the goal of eliminating all pollutants by 1985. The Clean Water Act left administration of water quality standards to the states, which were required to list damaged rivers and calculate total maximum daily loads (TMDLs), or the maximum amount of pollution they could withstand and still be drinkable, swimmable, and fishable. If any state failed to develop adequate standards, Congress gave the EPA the authority to step in and impose its own standards.

Appropriately, the Clean Water Act has been called one of the greatest successes in environmental law.[20] Our rivers no longer burn. And there are no more open sewers dumping crud directly into our rivers and streams. Over the last 40 years, the Clean Water Act has gone a long way toward reducing the amount of point source pollution discharged into U.S. waterways.

However, one of the loopholes of the Clean Water Act is that it does little to address nonpoint source pollution, such as sediments, pesticides, herbicides, fertilizers, and heavy metals that run off from farm fields into rivers. Although the Clean Water Act requires states to develop plans to

Figure 2.4. Polluted by decades of industrial waste, Ohio's Cuyahoga River caught fire multiple times, shocking the nation and spurring clean water reforms. *(Credit: Cleveland Plain Dealer, 1952. Obtained from Cleveland State University Library.)*

control these sources of pollution, it includes no sanctions for failing to do so. As a result, most states have failed to take the tough measures needed to address these problems. As the EPA itself has acknowledged, "Without a clear understanding of how to minimize pollution from . . . nonpoint sources, state and local organizations will be unable to develop strategies to protect their water resources."[21]

With the lack of strong controls, nonpoint source pollution has since emerged as the nation's largest water quality problem.[22] In fact, it is the main reason why 40 percent of U.S. rivers, lakes, and estuaries still aren't clean enough for swimming or fishing.[23] And the main culprit of nonpoint source pollution is agricultural runoff, which is largely responsible for degrading 60 percent of all damaged river miles and half of all lake acreage.[24]

Another shortcoming of the Clean Water Act is that it fails to effectively protect groundwater quality. Again, the authority here is dispersed between different laws and administrative agencies. For example, the Safe Drinking Water Act establishes regulations to prevent the contamination of well drinking water. The Comprehensive Environmental Response, Compensation, and Liability Act addresses groundwater contamination after the fact by requiring the control and cleanup of hazardous waste sites. And the U.S. Geological Survey collects groundwater quality data. What's more, some states have established programs to monitor groundwater quality, and in recent years the federal Subcommittee on Ground Water established a national groundwater monitoring framework, which is a large step in the right direction.[25] Yet no one law or agency is responsible for groundwater quality. Nor is the administration of groundwater quality tied to groundwater allocations, resulting in diluted laws that have failed to fully protect the quality of our groundwater.

A Twenty-First-Century Water Policy

If you've ever tried to remodel an old house, at some point you've probably run into problems. The replacement windows you needed may not have matched the standard sizes currently available. The old plumbing system may have used smaller-diameter pipes. Or perhaps the foundation and framing weren't built according to today's standards. After a while you probably realized the difficulty of maneuvering within the confines of the hand-me-down constructs you were given to work with. But you kept on tinkering anyway because the alternative would have meant tearing the whole thing down.

In the same way, the laws and regulations that govern water in the United States were developed during different eras with different sets of priorities without fully contemplating how they would appropriately toggle with all the others. They're a complex web that's no longer getting us the desired results, and we need to reset to address the priorities of our current era. The demand for water is outpacing the supply. Groundwater tables are shrinking faster than they're being recharged. And our watersheds are becoming increasingly damaged from agricultural runoff and other nonpoint source pollution. The reality is that the house we built no longer meets our needs. Yet we keep tinkering around the edges, making small modifications, when what we really need is a brand new house—one that allows us to tackle the twenty-first-century problems we face.

Ideally, a highly informed U.S. Congress ought to respond to the current era of water realities by passing a national act that integrates our water laws into a single, comprehensive policy. Surface water, groundwater, and water quality should be considered in unison, and their administration put under a single federal agency, to optimize the resource for environmental and economic gain. Given that water is arguably the most pressing resource crisis of the twenty-first century, it's disheartening to think that we lack a comprehensive policy for dealing with it. Instead, we're forced to sift through the clutter of various laws and regulations and do the best we can, which is absolutely the wrong mindset. As Winston Churchill so appropriately put it during World War II, "You have got to succeed in doing what is necessary."

Yet so far our legislators have failed to act, and with tight budgets and a bitterly divided and often dysfunctional Congress, any legislation at the federal level is improbable. The reality is that our water policies took centuries to become the complex jumble that they are today, and it won't be undone overnight. Complicating the matter is the fact that water crises tend to be regional in nature. And although we've witnessed water shocks in several areas around the United States, each is confined to a specific area of the country and hasn't affected the majority of Americans all at once, meaning the critical mass for legislation is unlikely. Drought in the Southwest doesn't necessarily motivate people in the Midwest to do anything. What's more, the slow-motion way that the environment inches toward disaster dissipates the energy around these issues. Even after the longest drought, it eventually rains again. And although the problem persists, the urgency to address it evaporates. The bottom line is that, although a national water act is the optimal solution, it's not likely to happen in the foreseeable future.

A Quantified Approach at Every Level

Although it's important to push for integrated laws at the national level, in the meantime, there's still a tremendous amount that government officials can do. For starters, they can mandate that everything be quantified to ensure that every taxpayer dollar spent renders a dollar in actual environmental benefit. In other words, they can require that outcomes be measurable and that twenty-first-century tools be used to monitor governments' success in achieving these goals. Given the scale of environmental problems we face, we can no longer afford to take a procedure-based approach. We need to demand strong outcomes, and technology can help get us there.

With today's technology, we can precisely identify the baseline from which we're starting, and we can do it in real time. And once the baseline's set, we can track what environmental improvements we make or what ground we're losing. And if we're losing ground, we can quickly make the needed adjustments, measuring again to make sure we've got it right this time. With tools such as geographic information systems (GIS) and light detection and ranging (LIDAR), for example, we can manage the impact of logging on watersheds via satellite or plane in real time and at less cost. Rather than physically driving to a site to survey the impact of a recent logging operation, we can use GIS and LIDAR to obtain precise images that allow government officials to make better decisions about the effect of these cutting operations on nearby watersheds. If a handful of trees are logged next to a river, they can see it. And if there's a landslide, they can see that too. They can also precisely measure the effect trees planted next to a river will have on water temperatures both today and as the trees grow larger over time.

Similarly, we can use data and analysis to focus spending on the projects that have the most impact. When restoring rivers, for example, it's no longer necessary for government agencies to work with every landowner who borders a watershed, as many governments do today at substantial cost. Instead, governments can save time, money, and effort by targeting only the landowners who are creating the biggest problems for the river.

As shown in figure 2.5, we can combine a Google Earth platform with public data to visually pinpoint which specific landowners are sending the most agricultural runoff into our streams and rivers. We can then determine which specific conservation practices on which specific plots of land would keep the most fertilizer out of nearby streams and rivers. For instance, would it be better to plant cover crops to absorb the excess fertilizer, or would a buffer strip of trees do the trick? Using quantified conservation, we can instantly determine that planting cover crops would keep 324 pounds of phosphorus out of the stream, whereas the filter strips would keep 162 pounds from reaching the water, and so on. Armed with precise data such as this, we now know where the greatest potential lies, and administrators looking to get the greatest uplift for the least cost can strategically provide incentives to the highest-polluting farmers to update their land management practices. We all conceptually understand that each piece of land is different, but we now can understand precisely how it is different and which specific strategies will render the greatest benefit. Although filter strips have been used more widely, the fact is that cover crops can do more good in certain scenarios. Engaging this precision allows us to replace subjective preferences with objective results needed to hit environmental targets and manage budgets so we can extend every conservation dollar we spend.

With data and analysis, we can also make water withdrawals in places that have the least impact on the river system. For example, the Columbia River serves as a superhighway for salmon, yet it's the small tributaries that contain critical quantities of water where the salmon spawn and rear. By developing a relational equation to determine the actual salmon benefit for a bucket of water in the small stream versus a bucket of water from the main-stem Columbia, we could convert irrigators to a more reliable

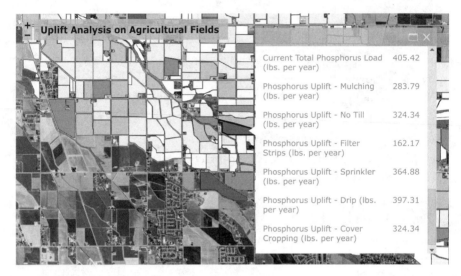

Uplift Analysis on Agricultural Fields	
Current Total Phosphorus Load (lbs. per year)	405.42
Phosphorus Uplift - Mulching (lbs. per year)	283.79
Phosphorus Uplift - No Till (lbs. per year)	324.34
Phosphorus Uplift - Filter Strips (lbs. per year)	162.17
Phosphorus Uplift - Sprinkler (lbs. per year)	364.88
Phosphorus Uplift - Drip (lbs. per year)	397.31
Phosphorus Uplift - Cover Cropping (lbs. per year)	324.34

Figure 2.5. Combining aerial images with existing data, we can identify the specific conservation projects on specific plots of land that will go furthest toward restoring our streams and rivers. *(Credit: The Freshwater Trust, using layers of publicly available data from the Snake River Basin.)*

water source while increasing the water in the small creeks where fish need it most. This would allow us to better manage where and when water withdrawals occur.

Of course, quantified conservation isn't limited to river restoration. Governments can use it to improve their results on almost any environmental project—whether it be to protect our forests, reduce the impacts of climate change, or increase wildlife habitat. And honestly, it's a complete no-brainer. Governments are already spending taxpayer money. At the very least, this money should be spent in a smart way. Just think if a pollster stopped the average person on the street and asked, "Do you want government to spend your money in such a way that they can precisely secure the intended results? Or would you prefer that government generally apply it in the direction of a problem with no worry of quantified results?" Even if the person on the street couldn't tell you how many senators there are in Congress, and even if they weren't able to place Washington, D.C. on a map, they'd still get that one right. Simply put, there's no logical reason why anyone wouldn't want to know what their public dollars are getting.

A Water Budget for Every Basin

With a quantified approach in place, we can then take the important step of developing a water budget for every river basin. Today, governments rarely measure how much water is available at the watershed level, let alone how much water needs to remain in-stream to maintain a healthy ecosystem. Yet the tools are available to do this. With the technology at our disposal, we can create accurate water budgets based on the total amount of surface and groundwater. We can also adjust these budgets in real time based on rainfall, snowpack levels, temperature, groundwater recharge rates, and other changing conditions.

In the same way that we budget for our personal needs based on the size of our paycheck, we can budget our water in a way that meets both economic and environmental needs only if we know how much total water we have in our bank account. Once we know how much water we've got, we can then create a flexible system in which we stretch our limited water in dry years while saving water and allowing for greater use in wet years.

As we establish water budgets, a key component involves setting a *minimum environmental flow*, or a baseline amount of water that needs to remain in the river. With so many of our rivers drying up every summer, minimum environmental flows are a must for any water budget. No matter how dry a year we're facing, our first priority should be to leave enough water in-stream for the long-term health of the river.

Beyond that, we can prorate the amount of water rights holders can withdraw based on their existing permits and how much water there is to work with. No doubt, prorating water will be controversial. But we have to do it if we're going to adequately manage water in an era of permanent water scarcity. Ironically, senior water rights holders who probably will protest the loudest over cutbacks to their water allocations could actually see their profits rise—if they find innovative ways to conserve water and then lease their excess water rights to other users who don't have enough. We've seen that happen both on an individual basis and in entire regions where rules were completely reset after droughts.

When I talk to landowners about these concepts, the conversation can get very heated. I've often heard people say, "You can't take away *my* water right."

To them I ask, "If I can make you more profitable by whatever means necessary, is that a better outcome? Or do you just want to philosophically use your water?"

In a quiet moment, some will admit that it's not really about the water. Rather, it's a cultural belief that more water equates to higher profits. But that's not always the case. In fact, it's often possible to make more money by converting to a higher-value crop that uses less water. If a farmer could earn $5 a bushel for a crop rather than 89 cents a bushel and use less water in the process, why wouldn't the farmer want to do that? That's exactly the kind of opportunity that more farmers would seize if we start to manage water as if it had value.

Similarly, we need to set a clear budget for the total amount of pollution our rivers can withstand while continuing to thrive and track progress in real time to hit and maintain targeted levels of water health. As discussed earlier, the Clean Water Act calls on each state to list its polluted watersheds and then work toward their cleanup by setting a TMDL. Those were to be completed by 1982, but several decades later, this homework remains undone for scores of river systems. Moreover, states continue to allow new discharges into rivers even after these assessments have been completed.[26] Whatever methods we have been using are not working. It's just a fact.

Not only does the situation need to be fixed, but we need to set TMDLs for entire river basins as they exist physically, not just state by state with small portions of rivers running through each and inconsistent parameters to track progress. Having TMDLs in place for every river basin would give us the information we need to better manage water pollution. Ultimately, it would enable us to balance a functioning economy with a functioning environment by allowing us to make the appropriate tradeoffs.

We've all heard the maxim, "You can't manage what you don't measure." Setting a water budget is a critical step toward ending our open-ended accounting system. It would allow us to track our environmental credits and debits so that we can make lasting improvements to the environment. It would encourage the conservation of clean freshwater. And it would pave the way for water trades to occur, ensuring that limited water supplies flow consistently to the most critical uses.

Using Existing Authority

Although a comprehensive water policy is the best way to manage our water, a quantified approach can also be driven by executive order, at both the federal and state levels. Such orders instruct agencies how to behave in organizing and managing their efforts toward a new outcome. Arguably the most tech-savvy of all presidents to date, President Obama has used his executive authority to inform government actions—including water policy—through the use of data and analysis. In a 2013 executive order, for example, he required federal agencies to take an inventory of their existing data, work to expand the maturity of their data, and make more of their data available to the public.[27] He has also issued executive orders requiring federal agencies to incorporate the highest level of scientific integrity into their decision making[28] and to examine existing regulations to determine whether they should be modified or streamlined "in light of changed circumstances, including the rise of new technologies."[29]

At the state level, executive orders can be the first step toward improving our watersheds as well. In drought-stricken California, for example, governor Jerry Brown in 2014 issued an executive order designed to strengthen the state's ability to manage water in drought conditions. The order directed state water authorities to expedite approvals of voluntary water transfers to assist farmers. It also called on businesses and individuals to conserve water and on homeowner associations to refrain from fining residents for limiting their lawn watering.[30]

Likewise, Oregon governor Kate Brown recently introduced a Clean Water Partnership that calls for water quality trends to be monitored and money spent on projects that are quantifiably shown to provide the greatest watershed improvements. The initiative calls for the acquisition of LIDAR data so that the state can develop baselines and prioritize investments. It also includes investment in an information technology infrastructure, so that water quality data can be shared between different agencies, and an executive order that aligns the various agencies so that they track progress in a standardized manner.

Executives at the federal and state levels can drive a lot of good. For example, they can introduce initiatives that require a quantified approach and then follow through to make sure there's a through-line from what's written on paper to what's actually implemented on the ground. Some examples include the following:

- Requiring that a water budget be established at the watershed level to properly allocate existing water between uses while leaving enough water in-stream for the long-term health of our rivers
- Compelling the inventory and modeling of the present state of rivers to determine the current baseline conditions and analyze which restoration projects have the most potential for environmental gain
- Authorizing the modeling of future scenarios to measure how different conditions will affect specific streams and rivers in the future
- Analyzing the quantitative success and shortcomings of past restoration efforts to inform course corrections
- Mandating that restoration projects be prioritized, both by river basin and by specific sites along each watershed that need the greatest improvement
- Requiring that all agencies working on water resource issues use the same set of "currencies" or "equivalencies" to measure and guide their efforts to track results in a standardized way
- Mandating the monitoring and evaluation of a river's conditions over a specific period of time once restoration efforts are completed

• Subjecting all these strategies to constant and real-time review through dynamic technology to ensure proper execution and necessary data-driven adjustments

By directing administrative agencies to follow requirements such as these, not just for rivers but for every environmental project, executives at the state and federal levels can ensure that money is spent far more effectively than it is today. Now that's an idea that should appeal to politicians on both sides of the aisle. The final design piece would include a planned "refresh" mechanism that will allow refinements that actually address emerging issues and realities on the ground and would disallow the legislative mischief that happens when environmental statutes have "planned sunsets."

In the case of water—and arguably with all environmental issues—maintaining the status quo is bad policy. The bottom line is that we need to fix more rivers faster, and we can. Quantified conservation offers a solid starting place, one that allows governments to determine whether we're getting the results we seek and to quickly change course if we're not. To close our open-ended accounting system, government officials must be willing to factor our impact on the environment into the overall equation. And they must be willing to lead despite the pains of change.

CHAPTER 3

Holding the Line Is Holding Back Environmentalism

MODERN ENVIRONMENTALISM is in need of a major overhaul. Despite some progress over the past generation, the majority of today's environmental groups have been using the same set of tactics that have been used since the environmental movement took hold in the 1960s and 1970s. The tools originally built for the job worked at one time, but today they're no longer keeping up with our evolving problems. As a result, environmentalists are obtaining an ever smaller return on their investment. Despite all the advocacy, legal victories, and public and private funding that have been funneled into environmental efforts, our waterways are in many ways far worse off today than they were a half century ago. As sad as it sounds, our combined efforts have added up to little more than a speed bump placed in front of a car with superb shock absorbers. To address the magnitude of the environmental challenges that confront us, we must revamp our approaches, and we must do so at a fundamental level. The time to patch, tinker, and improve parts here and there is done. We must change a much broader system, and that means a new environmentalism altogether.

My own disillusionment with mainstream environmentalism began early in my career at The Freshwater Trust, then called Oregon Trout. In 2001, just as I was getting my feet wet as president of the organization, we wrote an amicus brief[1] explaining the science behind what we saw as a dangerous lawsuit. Brought by the Alsea Valley Alliance, a coalition of property rights interests, the lawsuit sought to remove the coho salmon from its protected status under the Endangered Species Act. To do that, the alliance argued that the distinction between wild coho salmon and hatchery salmon was arbitrary. Counting both wild salmon and those that were artificially spawned would increase the total numbers of salmon in the rivers, pushing coastal coho salmon off the endangered species list. In the end, the U.S. District Court for the District of Oregon sided with the alliance, forcing the National Marine Fisheries Service (NMFS) to remove the coho salmon from the endangered species list throughout the state of Oregon.

Although my organization was first in line to appeal the decision, we decided against it. After reading the court's analysis and opinion, we concluded that it had been right on the law, which meant that our chances of winning any appeal were slim. We also concluded that the risks of failure were too great. If the U.S. Court of Appeals upheld the District Court's decision, the so-called fish is a fish ruling wouldn't apply just to Oregon. It would be extended to a much broader area that included Alaska, Hawaii, Washington, California, Montana, Idaho, Nevada, and Arizona, jeopardizing the survival of wild salmon throughout a vast portion of the American West.

We explained our conclusions to other environmental groups, pleading with them not to appeal the case. Yet despite our warnings, a coalition of environmental groups went ahead and filed an appeal. They were following a different agenda and needed the coho salmon on the endangered species list to file another lawsuit seeking stricter logging practices.

In the end, the appeal was dismissed, and we dodged a bullet.[2] Losing this case was a huge setback for all of us working to protect Pacific Northwest salmon, though, and I was both shocked and irked. In their zeal to save the forests, the coalition of environmental groups had literally been willing to sacrifice the fish. And, although the NMFS has since relisted the coho salmon as threatened,[3] for me the case symbolized how narrowly focused and ineffective environmentalism had become. It makes no sense to sacrifice one part of the environment for the benefit of another. If we are to be successful in the long term, we need to consider the big picture, and we need to establish outcomes that benefit the environment as a whole. Yet mainstream environmentalism has evolved into a series of tightly focused groups intent on defending narrowly defined causes—whether that be wildlife, fish, forests, or a wide range of other environmental topics. Rather than serving as stewards of the environment, we had turned into an array of sometimes competing special interest groups.

Speeding Up the Pace of Progress

When I was in my freshman year at Dartmouth College, my father died unexpectedly of a heart attack. His death shattered my world and showed me firsthand that life is short, which, in turn, made me all the more focused. I gave up my preoccupation with sports and girls and turned my attention almost exclusively to big ideas that could help make the world better. It was rare for someone from my small town in southern Illinois to be accepted into an Ivy League college, and I decided I didn't want to waste the opportunities I get.

As an adult, I've come to hate wasted potential. I believe that if you have the ability to accomplish something, you've got to make it happen—whatever it takes. Yet unfortunately, that's not what I'm seeing with the environmental movement. As I go about my work, I find it frustrating to observe so many super-smart people apply themselves harder and harder to the same old paradigm while getting such little return for their effort. Most of them have their hearts in the right place, but when I see how they approach their work, it's not taking them where they want to go.

In the same way that many of us take a step back once in a while to evaluate whether we're accomplishing our personal goals, it's important for the environmental movement to assess its effectiveness. What results have we been getting? Where have we been successful, and in what areas can we improve? Unfortunately, progress in the environmental world has been much too slow.

Consider how long it took to move from the concept of a land trust to a water trust. In 1951, The Nature Conservancy started as a nonprofit, popularizing the land trust model, in which nonprofits protect sensitive natural land areas by purchasing them and removing them from development. Yet it wasn't until 1993 that we took this concept and expanded it to protect rivers by forming the nation's first water trust. That's 42 whole years! Compare that with how far the Internet has come in just half that time. Twenty years ago, the Internet was this crazy new thing. Netscape's Marc Andreessen had just launched his "revolutionary" browser, Mosaic, and the Internet had just 16 million users. There was no Google. There was no Facebook or Twitter. There was no cloud computing. Today, of course, more than 3 billion people worldwide are connected to the Internet, and it's become an integral part of everyday life for 40 percent of people on the planet. If only the environmental movement moved at a fraction of this pace, our natural areas would be a lot better off for it.

The time has come for the environmental movement to admit that the results aren't adding up. Smart businesses remake themselves all the time—and for modern environmentalism, such an assessment is long overdue. By taking a step back to gauge where we've risen to meet the challenges before us, and where we've failed to adapt, we can adopt new tools better suited for the problems at hand. And that starts with evaluating how the modern environmental movement has evolved.

The Path of Modern Environmentalism

Until Rachel Carson's book *Silent Spring* was released in 1962, environmentalism had centered mainly on protecting wild spaces from development, and conservation focused on securing habitat for fish and game species. From John Muir to Henry David Thoreau to Aldo Leopold, environmentalists were focused on safeguarding land from human encroachment. What's more, environmentalism hadn't entered the mainstream.

Rather, it had been considered the domain of the privileged elite.[4] Carson's book brought environmentalism to a new level by shifting the conversation to pollution, in her case, the hazards of the pesticide DDT. Carson's best-selling book, combined with incidents such as the burning of the Cuyahoga River and the massive 1969 oil spill in Santa Barbara, California, focused the nation's attention on artificial chemicals and their impact on our water and air.

By 1970, when the first Earth Day celebration took place, the modern-day environmental movement had been born. The brainchild of U.S. senator Gaylord Nelson, the first Earth Day shifted the conversation from traditional land conservation to pollution and resource protection—specifically the fear of cancer and other diseases caused by toxic substances in our water and air. The event drew 20 million Americans, who organized protests against the destruction of the environment in rallies across the United States.[5] Modeled after anti–Vietnam War demonstrations called teach-ins, Earth Day was credited with spurring widespread public momentum for clean air and water. "It was a gamble," Nelson said later. "But it worked."[6]

What followed were the golden years of environmentalism, in which numerous pieces of landmark legislation were passed, including the Clean Water Act, the Clean Air Act, and the Endangered Species Act.[7] Viewed as a whole, these laws have been called "the greatest achievements of the nation's history."[8] The 1970s were also successful years for environmental litigation.[9] With a wide range of strong legislation on the books, the law was now on the side of the environment, creating the opening to sue government agencies that weren't adequately enforcing these laws. And with many judges sympathetic to environmental causes, environmentalists saw the judicial system as a powerful tool for social change, turning to the courts as a way to bring about environmental reform.

Yet the big environmental victories were short-lived. By the time Ronald Reagan became president in 1980, a powerful business lobby had descended on Washington, DC. A strong coalition of businesses led by the Business Roundtable systematically worked to attack environmental regulations, claiming that all the environmental policies had led to economic decline.[10] The Reagan administration had set an antiregulatory agenda that included weakening environmental protection measures, with James Watt, Reagan's appointee to the Department of the Interior, a prominent symbol of the administration's hostile attitude toward the environment. At the state and local levels, too, the environmental movement was starting to come under attack. The "wise use" movement[11]—one quite different from the "wise-use" interpretation of which Aldo Leopold spoke—rose from a network of loosely allied right-leaning grassroots and corporate interest groups formed to promote unfettered resource exploitation and hammer hard on environmental causes, eroding the position that pro-environment politics had enjoyed through the 1970s.

Responding to the antagonistic atmosphere, the environmental movement did several things to improve its position. It strengthened its adversarial stance through combative literature and a laser focus on advocacy designed to get legal results while building up memberships and budgets. At the same time, mainstream environmental groups responded by expanding their own presence in the nation's capital and eventually at the state and local levels. Many environmental groups revamped their management styles to take on a more corporate image, expanded and professionalized their staffs, and adopted an intense focus on lobbying at both the federal and state levels.

Although the environmental pendulum has swung back and forth over the years, both the political and litigation terrain have changed dramatically since the 1970s. The conservative wing of the Republican Party has grown substantially, giving environmental activists less room to maneuver inside the Beltway.[12] Even the legislative branch, which environmental groups had historically counted on for their success, could no longer be depended upon. As Christopher J. Bosso and Deborah Lynn Guber explain in their article "Maintaining Presence: Environmental Advocacy and the Permanent Campaign," "The access and leverage has since evaporated starting with the shift of congressional control to Republicans in 1995. In the decade that followed, environmentalists found themselves essentially excluded from the innermost circles of House decision making, and watched as their legislative proposals disappeared from the agenda."[13]

In addition, Republican dominance of the presidency from the 1980s through the first decade of the twenty-first century helped create a conservative orientation of the federal judicial branch to environmental and regulatory issues. The result has been that lawsuits that once shaped environmental policy have become "little more than narrow-gauge tools for forcing over-burdened regulatory agencies to adhere to the letter of the existing law," as Bosso and Guber put it.[14]

Ironically, environmentalism has become less effective as a movement at the same time as support for environmental causes has skyrocketed. There are currently more than 15,000 environmental and animal public charities in the United States, compared with just a few hundred in 1970.[15] Membership in U.S. environmental organizations has more than tripled, from 5 million in 1981[16] to 16 million today. [17] In addition, revenues have climbed to nearly $15 billion,[18] and Earth Day has grown from 20 million participants at its launch in 1970 to a worldwide movement that attracts more than 1 billion people each year.[19]

There's no question that growing membership bases, clean water legislation, and many of the lawsuits that followed have all advanced the cause. Thanks to the efforts of a wide range of people, our water is cleaner and more drinkable in many places around the country than it would have been had the movement not emerged. Yet despite a lot of hard work and a lot of important victories, many of our streams and rivers are headed in the wrong direction at a time when climate change and population growth are putting unprecedented strain on freshwater. As has often been said, the proof is in the pudding.

To assess the state of our freshwater, and indeed the environment as a whole, one need look no further than what's happening on the ground. And unfortunately, every major indicator of environmental health has been heading in the wrong direction. As a group of 1,700 of the world's leading scientists warned more than 2 decades ago,

Human beings and the natural world are on a collision course. . . . We are fast approaching many of the earth's limits. Current economic practices which damage the environment, in both developed and underdeveloped nations, cannot be continued without the risk that vital global systems will be damaged beyond repair.[20]

The Failure to Adapt

So where have environmentalists gone astray? In a nutshell, today's environmental movement keeps recycling the same old playbook from the last 45 years, even as the situation has changed, and these tools are no longer a match for our escalating problems. Both the political and litigation realities we face today are far different from the ones environmentalists confronted at the dawn of the modern environmental movement. What's more, the environmental issues themselves are different. From climate change to nonpoint source pollution to the effects of population growth, the fact is that the environmental problems we face today are far more massive and entrenched than the ones environmentalists confronted in the 1970s.

Yet rather than coming to terms with these critical facts, environmental groups continue to press ahead with these same advocacy and litigation tools and tactics, continuing to hone and expand them without taking a step back to evaluate whether they're still working. Although the focus on advocacy and litigation initially resulted in large wins for the environment, these tactics are no longer as effective as they once were. As a result, while we're winning some battles on paper, we're losing the war on the ground.[21]

Over the years, some in the environmental community have been sounding the alarm bells. For example, in a famous December 2004 speech, "Is Environmentalism Dead?," former Sierra Club president Adam Werbach argued that environmentalism had reached the limits of its effectiveness. "Nobody enjoys an autopsy, and yet its value to life is indisputable," he said. The speech followed an October 2004 essay, "The

Death of Environmentalism," in which Michael Shellenberger and Ted Norhaus argued that environmentalism has stagnated as a vital force for cultural and political change: "We have become convinced that modern environmentalism, with all of its unexamined assumptions, outdated concepts and exhausted strategies must die so that something new can live."[22]

So what must die so that something else can live? One is the assumption that environmentalism is continuing to have a big impact. Another is the almost exclusive focus on advocacy and litigation. As the environmental movement clings to its unexamined assumptions and exhausted strategies, it is gradually slipping into irrelevance.

Wrapped Up in Advocacy

One of the main focuses of the modern environmental movement has been on advocacy, including public education, lobbying, and fundraising. "Grand Canyon Under Siege," reads one environmental website.[23] "Victory! Coal Export Permit Denied!" reads another.[24] More often than not, environmental advocacy efforts are focused on stopping further damage from happening—whether it's the destruction that should be avoided, the extraction that should not occur, or the species that should not go extinct. One problem with this approach is that it isn't bold enough. With so many rivers and streams in jeopardy, it's not enough to hold the line. Holding the line means losing.[25] If we are serious about addressing our freshwater problems, we must set bolder outcomes that include turning around the health of streams and rivers that have already been damaged.

A second problem with this approach is that the big wall of "no" is no longer effective. In 1970, when the first Earth Day took place, Americans were just starting to understand environmental degradation. Awareness of the dangerous effects of toxic chemicals in our water and air were just beginning to seep into the public consciousness. Fearing the consequences to their health, Americans were outraged and demanded change.

The greens seized on the obviously effective tactics and ran the same play for years, and the result is that people are now overloaded with information about environmental problems, and much of the advocacy is backfiring. Consider this: Each year, hundreds of millions of fundraising pitches are mailed out by environmental groups. That equates to enough envelopes, stationery, decals, bumper stickers, calendars, and personal address labels to circle the earth more than two times.[26] Americans are being bombarded with pleas to help the environment, and the net effect is that it's creating a culture of despair. "What you get in your mailbox is a never-ending stream of crisis-related shrill material designed to evoke emotions so you will sit down and write a check," said Daniel Beard, former chief operating officer of the Audubon Society. "I think it's a slow walk down a dead-end road. You reach the point where people get turned off."[27]

Indeed, with so many crises competing for their attention, Americans are suffering from environmental warning fatigue. Not surprisingly, a recent poll found that citizens around the world, including those in the United States, are less worried about the environment at a time when environmental problems are growing in severity.[28] With so much emphasis on the problems and so little on the solutions, many Americans have simply checked out.

As part of their advocacy efforts, mainstream environmental groups continue to devote a significant portion of their efforts to government lobbying, even as the actual gains continue to diminish. In the late 1970s, when the environmental movement was first gaining power, environmentalists were actually part of the revolving door in Washington, DC. It wasn't uncommon for members of environmental groups to hold prominent government positions. From Joseph Browder of the Environmental Policy Center to James Gustave Speth and John Bryson of the Natural Resources Defense Council, environmental activists secured positions at the federal and state levels, where they worked front and center to formulate environmental policies.[29] Today, by contrast, many government positions go to industry executives, demonstrating just how much political ground the environmental movement has lost.

The modern environmental movement also uses advocacy as a fundraising tool. Today's large environmental groups typically rely on small donations from a large number of members. However, as environmental groups have grown in size, they've joined the mainstream, taking on corporate-style offices while adding large cadres of staff that include lawyers, lobbyists, scientists, economists, organizers, fundraisers, publicists, and political operatives. And, although both their memberships and revenue have grown too, a large chunk of these funds aren't used to protect the environment. Instead, they're used to pay for overhead and to fund future fundraising campaigns. As Pulitzer Prize–winning journalist Tom Knudson pointed out in his excellent *Sacramento Bee* series, "Environment Inc.," a large portion of today's mainstream environmental groups spend more than 35 percent of their revenue on future fundraising campaigns, exceeding the recommendations of philanthropy watchdogs.[30] "In truth, what the environmental community has become is a money machine," author Alfred Runte told *The Sacramento Bee.* "We have come to the point where we keep score by the almighty dollar. And we need to start keeping score by the health of the environment."[31]

Keeping score by the health of the environment, indeed! If we did that, we'd quickly realize that our efforts are no longer adding up. As Adam Werbach, the former Sierra Club president, has said, "Most of my colleagues have committed to arguing better, yelling louder and organizing more people. But no amount of public relations or grassroots organizing will move problems like global warming up the list of issues Americans worry about."[32]

Mired Down in Litigation

Another area in which the environmental movement has devoted its efforts is litigation. Over the past half century, many national environmental groups have focused a good portion of both their energy and their budgets on lawsuits intended to force government agencies to comply with legislation that requires them to protect the environment. The complexities have increased to get every last little inch of gain, but the wins on paper have plateaued in terms of actual results. The planet receives little gain from our self-proclaimed "big wins." We're fooling ourselves.

Citizen suits began cropping up in the early 1970s after the passage of the Endangered Species and Clean Water acts, which gave ordinary citizens and environmental groups the power to hold government agencies accountable in the courtroom when they failed to enforce the law or properly follow mandated procedural safeguards in their decision making.[33] Some of these suits have significantly helped to protect the environment. Thanks to one citizen suit, for example, the Natural Resources Defense Council successfully forced the U.S. Bureau of Reclamation to restore water to the San Joaquin River. Thanks to another such suit, the northern spotted owl was successfully listed as a "threatened" species, dramatically reducing poorly planned logging in the Pacific Northwest.

Yet, over time, litigation has gotten an ever smaller return on its investment. "Creative litigation," as it's known in green legal circles, is the process of using protections afforded by one environmental law or court decision to gain enough of a handhold in another area of environmental interest to extend our reach and ability to reduce bad actions. Although creative litigation has been responsible for some important wins, increasingly these suits take decades to resolve, often with no tangible, on-the-ground results after the win is concluded.

Ironically, some of these legal actions have actually been hurting the environment rather than helping it by tying up public agencies that are already tied up in legal knots. In the 1990s, for example, the government paid $31.6 million in attorney fees for more than 430 environmental lawsuits brought against federal agencies.[34] In the scheme of things, that's not a huge number, but it's not the only cost. The hours and resources spent by governments dealing with these lawsuits surely dwarf this figure. Moreover, the defensive mental state into which we drive agency staffers puts a freeze on anything that looks like collaborative problem solving. The relationship becomes abusive. Essentially, we hit the agencies if they move too fast, we hit them if they move too slow, and we hit them if they don't move at all.

With so many lawsuits, government biologists have been spending more time defending themselves from legal action than on field conservation work, ignoring species that need protection. What's more, these lawsuits are forcing federal agencies to spend their limited budgets on attorney fees rather than on efforts to protect the environment. "We've filed our share of lawsuits and I'm proud of a lot of them," said Dan Taylor, executive director of the California chapter of the National Audubon Society. "But I do think litigation is overused. In many cases, it's hard to identify what the strategic goal is."[35]

An Us versus Them Mentality

Not only has the relentless focus on advocacy and litigation failed to bring about the results we need, but it has resulted in a brutal us versus them mentality in which those who conserve resources are pitted against those who extract them. On one side, agriculture and industry believe environmentalists are out of touch and out to end their way of life. On the other, environmentalists distrust those who work in agriculture and industry, viewing them as shortsighted and selfish.

Over time, the divisions have become more entrenched. Take the U.S. Chamber of Commerce, which directly attacked environmentalists in its 2004 booklet *Top 10 Environmental Myths*:

> Myth 4: All environmentalists are motivated by altruistic concern for the planet.
>
> Fact: Environmentalists hype scare tactics to raise money. . . .
>
> Myth 5: Environmentalists are all penniless college students, backed by overwhelming scientific opinion.
>
> Fact: Environmental groups have enormous wealth, cry wolf to raise billions of dollars, and their most serious claims have been proven wrong.[36]

Language such as this seeks to demonize rather than move the discussion forward. But environmentalists have also contributed to the polarized environment. Consider Earth First's motto, "No Compromise in Defense of Mother Earth."[37] And even mainstream environmental groups lash out at agriculturalists and industry rather than focusing on the facts. "Last week, we succeeded in bringing down a Goliath . . . none other than

Exxon Mobil," reads one fundraising e-mail. "The people have prevailed over Big Oil!" reads another.

The underlying problem is that there's no common ground between the players, and yelling like this doesn't really help. Whereas an agricultural producer translates land into bushels of grain or pounds of beef, an environmental advocate sees land as a place of beauty. There is no lingua franca between the two and thus no starting point from which the various players can communicate.

When I talk to environmentalists about using monetary incentives to motivate landowners to conserve water, I sometimes hear, "But the greedy bastards should be doing it anyway." And I think, "You're protesting grazing subsidies, yet here you are wearing leather shoes." Unless you're living in a mud hut reading this book by a candle, the fact is you probably have a massive footprint well beyond what you realize. We all do, myself included. Yet many environmentalists take an almost religious attitude toward these issues, as if they're the good guys, and businesspeople are the bad guys who need to be conquered.

In the same way, environmentalists love to throw stones at large corporations rather than work with them to change their practices. Just think about what happened to former Sierra Club president Adam Werbach when he decided to serve as an environmental consultant to Walmart. Old friends wouldn't speak to him. Former colleagues accused him of selling out. He even began receiving physical threats.[38] Yet sometimes success requires motivating the wrong people to do the right things. The Walmarts of the world, and the customers who buy their products, have a tremendous impact on the planet. If environmentalists can persuade multinational corporations such as Walmart to lessen their environmental impact, they haven't created the perfect world, but it's still a huge win for the environment.

No doubt we need a new system, but we cannot pretend for a second that we are separate from it. The reality is that we need agriculture and industry, *and* we need a healthy environment. In this era, these are not "nice-to-haves." These are "gotta haves." There's no getting around the fact that humans will use resources—we always have throughout our existence. We need to eat, so we need farms. We also need to eat in the long term, so we need agriculture to operate within a healthy ecosystem. Likewise, we need clothing, housing, transportation, and a range of other daily necessities, so we need industry. These needs won't go away tomorrow—and neither will Walmart. In the meantime, the hardened philosophical divisions are getting us nowhere.

The Need for a New Approach

The time is ripe for a new approach to the environment, one that fosters cooperation rather than clashes and one that harnesses the latest science and technology to address the magnitude of the problems in front of us. No matter how urgent, humans end up burning out on problems. Americans are tired of the us versus them mentality. They're tired of endless crises and seemingly insurmountable problems. What they want are creative solutions that result in measurable gains for the environment while taking into account the needs of the economy. As Dan Taylor, executive director of the California chapter of the National Audubon Society, put it, "We've effectively sold the idea that the world is screwed up. What people are looking for now are durable solutions on how to make it better."[39]

Broadening Our Toolbox

I can talk to industry executives all day about why they need to protect our rivers, and they say, "Yes, but—," and I don't end up with a commitment. But when I say, "You're going to get the compliance you need with the Clean Water Act and save money in the process," that's it. Discussion over. I always build in a big gain for the environment, but that's a detail. Explaining to people why they should do the right thing for the environment generally doesn't work, but when I give them an incentive that works to their benefit, their eyes begin to light up. As author and inventor Buckminster Fuller once said, "If you want to teach people a new way of thinking, don't bother trying to teach them. Instead, give them a tool, the use of which will lead to new ways of thinking." As environmentalists, we have additional tools in our toolbox. We just need to put them to use. As a movement, we need to establish more ambitious outcomes, design new tools for the problems at hand, and then use data and analysis to inform the success of our work.

Consider the growing number of animals faced with extinction both in the United States and around the world. By some estimates, as many as 30,000 species are being driven to extinction each year.[40] Through advocacy, environmentalists have done a great job calling attention to this problem. And through numerous lawsuits, they've successfully gotten species listed as "endangered" or "threatened," which affords them some protection. Yet litigation and advocacy have gotten us only so far. To date, only 1 percent of species on the endangered species list have actually recovered to the point where they can be removed from the list.[41] With the world's plants and animals going extinct at a rate 1,000 to 10,000 times faster than they did before humans came along, adding more species to the endangered species list isn't enough.

Our ultimate goal must be to help these species recover to the point where the protections of the act are no longer needed for their survival. Yet here the environmental community as a whole has been much less effective. Interestingly, the incentive in some cases has been to keep endangered species on the list because it provides the legal handhold needed to sue on almost any habitat issue in any region. (Under the working theory of creative litigation, that's how to slowly build a bulwark against extinction.) Without another model to go about protecting the parts of nature we care about, we're constrained by our hand-me-down tools. Unfortunately, the wins on paper are no longer translating to wins on the ground.

The good news is we have the tools at our disposal to make gains and help fend off extinctions. With today's science and technology, we can perform analyses to identify what areas are the most cost-effective to protect. And with this information in hand, we can engage in restoration work in a highly targeted way, using data and analysis to measure our success, freeing up resources to do even more.

One environmental group that's doing exactly that is SavingSpecies, a nonprofit aimed at restoring the world's biodiversity.[42] Using the best available science and technology, the organization is working to identify exactly what parcels of land need to be restored and reconnected in order to save specific species. It then engages the local community, helping organizations to raise the funds needed to restore habitats that will prevent the most species extinctions. "Often it requires very modest purchases of land," said Stuart Pimm, the organization's chairman. "East of Rio de Janeiro, for instance, it only required a few hundred hectares to stitch together a piece of forest that's 8,000 hectares. And in doing so, [we've achieved] huge impact on a charismatic monkey called the golden lion tamarin. That came about from using very focused science to identify the key areas, understand the key processes, and going in there and working with local communities."[43]

Note that SavingSpecies is attempting to accomplish its work not to the exclusion of human activity but in balance with it. In the same way, environmentalists can use data and analytics to prioritize their efforts to restore the environment, complete the work in a cost-effective way, and then measure their results. Rather than tracking wins and losses in the courtroom, we must track our results ecologically.

Likewise, Conservation International has teamed up with Hewlett-Packard (HP) to improve the accuracy and speed of data analysis tracking threats to wildlife in tropical forests. The initiative, called "HP Earth Insights," generates trend data on endangered animals using near real-time data analytics with big data technology supplied by HP. To gather data, the two organizations have set up about 1,000 camera traps and climate sensors in countries ranging from Brazil to Uganda and Indonesia. The photos and sensors monitor everything from vegetation to precipitation to carbon stocks, and the data they generate are then analyzed using complex statistical methods.[44]

Using this technology, the organizations have been able to quickly narrow down species that are dwindling in population so that quick action can be taken. For example, early results have shown that 33 of 275 species being monitored have significantly decreased in numbers, giving environmentalists the data they need to better focus their efforts.[45] "HP Earth Insights is transforming environmental science," said Peter Seligmann, chairman and chief executive officer of Conservation International. "Until now, the right data, the technology and scale have been noticeably missing from our field. What once took a team of scientists weeks, months or more to analyze can now be done by a single person in hours."[46]

Bridging Economic and Environmental Interests

Not only can science and technology help us prioritize our efforts and evaluate our work, but it can bridge the gap between economic and environmental interests by giving them a common language with which to make tradeoffs. No longer are environmentalists limited to explaining the value of nature in terms of the beauty it offers. We now have the tools to accurately describe the environment according to the services it provides to humans and to broader ecosystem resilience, and we can quantify these services in a way that can ultimately assign them a dollar value in order to make the right tradeoff (figure 3.1).

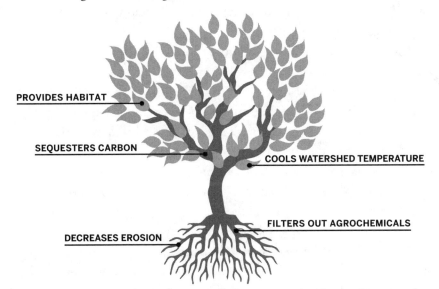

Figure 3.1. Trees provide a wide range of ecosystem services that can be assigned a precise value. *(Credit: The Freshwater Trust.)*

Interest in quantifying "ecosystem services" has been growing ever since Robert Costanza and a team of researchers added up the value of the world's ecosystem services in 1997 and published their results in the journal *Nature*. Costanza and his team put the value of services provided by ecosystems at $33 trillion annually—nearly twice the size of the gross national product of all countries in the world.[47] The concept became even more popular in 2005 when the United Nations published a report that grouped ecosystem services into four types of benefits: *provisioning*, such as supplying water, food, and wood; *regulating*, such as controlling erosion or purifying water; *supporting*, such as pollinating crops or creating soil; and *cultural*, such as offering spiritual and recreational benefits.[48]

By quantifying the benefits of ecosystem services and assigning them a monetary value, environmental and economic interests have a common language that all parties can understand, allowing them to more accurately weigh the tradeoffs when making decisions. For example, after Congress amended the Safe Drinking Water Act, New York City faced the prospect of installing an expensive artificial water filtration plant that would cost $6 to $8 billion, with $300 to $500 million in annual maintenance costs (figure 3.2).[49]

Instead, the city decided to improve its water quality by implementing a watershed protection program that involved buying up tracts of land to serve as buffers. Not only could they complete the alternative plan at a fraction of the cost, but it guaranteed better water quality indefinitely. Tradeoffs such as this are winners, and we must work to make them the norm rather than the anecdote.

With today's ability to quantify ecosystem services and assign them a monetary value, conservationists can better guide restoration investment and action. With the ability to measure the exact benefits that nature provides, we can target what ecosystem services must be retained or added to offset the impacts from industry. By pivoting from scare tactics to focusing effort where the environment needs it most, we can also determine which restoration efforts will have the greatest impact, targeting our limited resources to the most important projects. With the magnitude

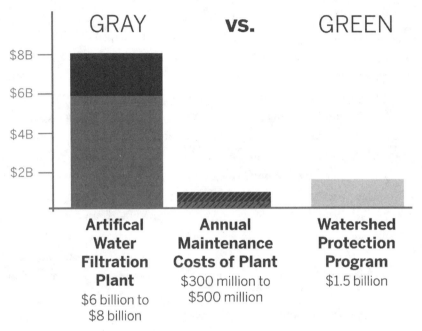

Figure 3.2. By choosing green over gray infrastructure, New York City saved money while protecting its water quality indefinitely. *(Credit: The Freshwater Trust, using data from Robert Costanza, Ralph D'Arge, Rudolph de Groot, et al., "The Value of the World's Ecosystem Services and Natural Capital," Nature 387, May 15, 1997, pp. 253–60.)*

of freshwater challenges in front of us, the quantification of ecosystem services provides environmentalists with a powerful tool to make every effort efficient and effective.

Putting Environmental Funding to More Effective Use

When it comes to restoring natural areas, quantified conservation also offers us the ability to maximize every dollar spent by tying investments to measurable gains for the environment. Although there will always be the need for some advocacy and some litigation, imagine that a sizable chunk of the $15 billion in annual U.S. environmental nonprofit revenue were instead spent on restoration programs tied to measurable improvements for the environment. Certainly, we would see many more gains for the environment than what's currently taking place.

With the ability to prioritize the most important projects and tie investments directly to results, environmental organizations could allow their members to earmark their donations for specific initiatives while informing them of the exact benefits achieved. In addition, environmental groups could form alliances with agriculture and industry to complete high-priority restoration projects, tapping into the growing corporate responsibility budgets to pay for them.

To determine how to spend one's efforts, Google's Larry Page suggests asking the threshold question, "Am I working on something that can change the world?" Environmentalists across the board can appropriately answer "yes" to that question. But that's not enough. We must also ask ourselves, "Am I working on that something in a way that actually *will* change the world?" If we are honest with ourselves, the answer to that question is almost universally "no." It's admirable to think boldly, but we must follow through with effective action. Quantified conservation provides the way forward. It can transform environmentalism into a more powerful force for protecting our rivers and streams while creating a more durable prosperity within the limits of the biosphere.

Environmentalists can and should lead the way. Yet the principles that underlie this new approach require that all of us work together. As we will see in the next chapter, maintaining a thriving agricultural industry during a time of permanent water scarcity requires that agricultural interests take a seat at this new table as well.

CHAPTER 4

Real Cowboys Fix Rivers

Faced with inadequate water supplies as California experienced its third driest year on record, San Joaquin farmer Barat Bisabri recently watched his water allocation drop to zero, forcing him to purchase water at high prices while taking acreage out of production.

Managing partner of Shiraz Ranch in the San Joaquin Valley, Bisabri gave his pistachio trees less water, which reduced his yield by 25 to 30 percent. He was also forced to cut 85 acres of mandarin trees out of production, literally letting 15 percent of his citrus trees dry up and die. "When I calculated the amount of water we'd potentially have, there was no way I was going to have enough water for the whole acreage," Bisabri said in an interview. "The problem with mandarins, which is the majority of my crop, is that you have to have the right size. And if you don't give them enough water . . . basically it's not going to be worth it to pick them because they're going to be too small."

At the same time that he cut production, Bisabri kept other parts of his fruit and nut trees in operation by purchasing $1.2 million worth of water—ten times what he spent on water purchases just 2 years earlier. He had hoped to buy even more water from a nearby property owner who had been pumping and selling tens of thousands of gallons of well water, but the opportunity withered after the property owner's neighbors complained all the pumping was diminishing their own well water supplies.

Bisabri hopes that the situation will turn around next year and that he'll be able to buy the property owner's well water. Yet he admits he's worried about the future. "With a little bit of rain, I think we're going to be okay," he said. "But if it doesn't rain, I don't know what we're going to do."

Bisabri is among hundreds of farmers who are feeling the effects of water scarcity in California's Central Valley. A large, flat area that actually encompasses two valleys—Sacramento to the north and San Joaquin to the south—the 450-mile-long Central Valley is one of the largest and most productive agricultural regions in the world. The region grows more than 250 different crops[1] and produces nearly half of the nation's fruit, vegetables, and nuts, as well as four fifths of the world's almonds.[2]

Even before the drought, the persistent overuse of water combined with explosive population growth had taxed limited water supplies, causing rivers to run dry, dead fish to accumulate near the water pumps, and chronic water shortages. Historically, the San Joaquin Valley had been a water-poor region of the state. To transform the semiarid desert environment into productive farmland, the U.S. Bureau of Reclamation in the 1930s took water from rivers in the northern part of the state and transported it to the south via a series of canals, aqueducts, and pumping plants.

Initially, the so-called Central Valley Project spurred California's booming agricultural industry. Yet, over time, the availability of water hasn't kept up with the pace of agriculture. In the meantime, the project changed an entire ecosystem, lowering river flows in northern California, destroying salmon runs, devastating marshes and wetlands, and increasing the salinity of watersheds. Compounding the problem is the fact that California's population has skyrocketed from less than 6 million in 1930 to about 38 million today, increasing competition for water from municipal and industrial users.

Over the past few years, California's drought has exacerbated what was already a critical problem. The drought is responsible for the greatest water loss ever seen in California agriculture, with river water for Central

Valley farms reduced by roughly one third. In 2014 alone, the drought was expected to cost the farming industry more than $1 billion in crop, dairy, and livestock losses and cause more than 17,000 workers to lose their jobs, according to a report from scientists and economists at the University of California, Davis. With nearly 80 percent of the state experiencing "extreme" or "exceptional" drought, the farming industry has come under increasing risk.[3]

Like Bisabri, farmers in parts of the Central Valley have fallowed some of their land to keep other parts alive.[4] Others are pumping groundwater at unprecedented rates, causing water levels to drop below the reach of irrigation pumps.[5] Chronic overpumping has depleted ground-fed surface streams[6] while causing some municipal wells to run dry.[7] In the meantime, water contamination from agricultural runoff in the valley has increased fivefold in the last 4 decades, angering residents whose nitrogen-laced well water has been rendered undrinkable.[8]

Today, unemployment in some towns of the San Joaquin Valley has approached 40 percent, with the worker motto, "No water, no work." Ironically, long food lines have become daily occurrences in a region that produces more food than anywhere else in the country.[9]

Responding to the crisis, the California legislature in 2014 tried to limit groundwater pumping for the first time. The new laws give local agencies 5 to 7 years to develop their plans and until 2040 to implement them, leading some to criticize the legislation as "too little, too late."[10] In the meantime, farmers and ranchers continue to withdraw water as if it were an unlimited bank account. "We're acting like the super-rich who have so much money they don't need to balance their checkbook," said Richard Howitt, coauthor of the University of California, Davis report.[11]

U.S. Agriculture at a Crossroads

The Central Valley is not alone in its struggle for water. If we continue down the current path, agriculture across the United States will not have the supply of clean freshwater needed to feed a growing population. Nor will it continue to enjoy its status as the world's largest agricultural exporter.[12] Today, irrigation is the most significant use of water in the United States, accounting for 80 to 90 percent of the nation's consumptive water use.[13] Overall, freshwater withdrawals in the United States have nearly doubled since 1950,[14] and with population in the United States expected to grow by another 25 percent by 2050, the agriculture industry will confront growing instability in both the availability and the cost of water. Combine that with the increasing occurrence of drought and flooding caused by climate change and growing competition for water from municipalities and industry, and the future of water for U.S. agriculture is a cause for grave concern.

Internationally, the rate of agricultural production growth has already been slowing because of both shrinking water supplies and the increased competition for land, according to a recent United Nations report, *The State of the World's Land and Water Resources for Food and Agriculture.* "Water scarcity is growing," the report said. "Salinization and pollution of water courses and bodies, and degradation of water-related ecosystems are rising. In many large rivers, only 5 percent of former water volumes remain in-stream, and some rivers . . . no longer reach the sea year-round."[15]

Unfortunately, the slowdown has been occurring at a time when food production needs to increase dramatically to accommodate rising demand. Global population is expected to grow by another 2 to 3 billion people by 2050. At the same time, worldwide demand for food over the next few decades will increase by a staggering 70 percent.[16]

Producing 70 percent more food than we do today will be a daunting challenge for a system that's already at or near capacity. Not counting Greenland or Antarctica, farms already cover nearly 40 percent of Earth's land surface, most of it the best arable land. And flows of nitrogen and phosphorus through the environment have doubled since 1960, creating dead zones at the mouths of many of the world's major rivers.[17]

Given all of this, using unlimited amounts of water, fertilizers, herbicides, and pesticides on unlimited swaths of land is no longer an option. Instead, we must feed the world's booming population by growing more food on virtually the same amount of land, with less water, in a way that's environmentally sustainable. And to achieve that goal, modern agriculture as we know it needs to undergo a significant redesign.[18]

Victims of Our Own Success

Over the last century, U.S. agriculture has made phenomenal gains. In 1880, nearly half of all working people in the United States were farmers.[19] Today, just 3 percent of the labor force produces all of our food and fiber while supplying about 10 percent of international consumption.[20] In terms of labor, it now takes just a fraction of the time that it once did to grow most crops. Today, for example, only 6 hours of human labor are needed to grow 100 bushels of wheat, compared with 147 hours in 1900. Likewise, it now takes just 5 hours to grow 1 bale of cotton, compared with 248 hours in 1900.[21]

Not only has labor efficiency increased, but so has the yield per acre. For example, the amount of corn farmers can grow has skyrocketed from 25 bushels per acre in 1900 to more than 120 bushels today. Similarly, the amount of cotton grown per acre has nearly tripled from about 200 pounds in the early 1900s[22] to 600 pounds today.[23]

Many of these efficiencies can be attributed to the Green Revolution that started after World War II, when the additional demand for food led to the rapid intensification of agriculture. Indeed, in the quarter century from 1950 to 1975, agricultural productivity advanced more quickly than at any other time in American history.[24] Thanks to the Green Revolution, farming today is dominated by the use of sophisticated equipment; synthetic pesticides, herbicides, and fertilizers; and hybrid strains of plants and animals. The benefits have been many. Electronic monitoring systems help farmers plant seeds at the necessary depth and spacing, and self-propelled harvesters have mechanized the harvesting of nearly every crop, with some even cleaning them and packing them into boxes. Fertilizers such as nitrogen have made it possible to grow crops on the same land year after year. Pesticides and herbicides have eliminated the need to cultivate row crops while dramatically increasing the yield per acre. And plant and animal breeding has led to higher crop yields, greater milk and meat output, and more disease-resistant crops. The Green Revolution has exponentially increased the amount of food available worldwide, allowing population to soar.

Yet, in many ways, the Green Revolution is a victim of its own success. Largely because of these farming methods, world population has doubled since 1960, putting greater pressure on both the supply and the quality of water. In the United States, pesticides can be found in nearly all streams in both agricultural and rural areas that have been studied by the U.S. Geological Survey. In addition, pesticides have been found in half of all shallow wells and one third of deeper wells that have been sampled.[25] Altogether, 42 percent of the nation's stream length is in poor biological condition, and 25 percent is in only fair biological condition, largely because of agrochemicals.[26]

In terms of our water quantity, industrialized agriculture has reduced the total amount of available surface and ground water. More than half of all crops west of the United States' 100th meridian use irrigated water from dammed rivers and aquifers for their production. And by some estimates, about one quarter of water used for irrigation isn't replenished on an annual basis.[27] Several factors, including rising population, economic growth, and increasing energy demands, are expected to further strain existing water supplies over the next several decades. At the same time, climate change—through warming temperatures, shifting precipitation patterns, and reduced snowpack—is expected to reduce water supplies and increase the demand for water, especially across the American West.[28]

Forty Cowboy Hats

Insufficient clean freshwater doesn't just hurt the environment; it also threatens the future of U.S. agriculture. As farmers in Central Valley, California have experienced firsthand, when the water runs out, agricultural production is put at tremendous risk. Perhaps as a reaction to the finger pointing of modern environmentalism, mainstream agriculture has done its own part to foster an us versus them mentality. The result has been a polarized atmosphere that has slowed the implementation of innovative solutions needed to maintain an adequate supply of clean freshwater, jeopardizing both the environment and the industry's future.

Although it is not universally the case, I've witnessed farmers' and ranchers' reluctance to change—and, in some cases, even to engage in discussions. One time when I was addressing a local watershed council about a pilot project designed to increase farming profits while saving water, I saw firsthand how the agricultural community can cling to an I've-always-done-it-this-way mentality.

The council was made up of citizens, resource agencies, and restoration professionals, with about forty farmers and ranchers, almost all of them wearing cowboy hats, looking on. After discussions lasting more than a year, for the umpteenth time, I reiterated the purpose of the project to the still suspicious crowd, who figured I was trying to steal their water and run them off the land that their forebears had settled. What I was actually trying to do was find an economically viable way for them to stay on their land while improving its health. But culture would not even let us get to the problem, much less the solution. The exchange went down like this:

I asked the sea of cowboy hats, "Who here would be interested in seeing if they could clear $30,000 rather than $5,000 a year, use less water, and turn that water back in-stream for fish—whether you love fish, hate fish, or couldn't give a rip about fish?"

Nobody raised a hand.

"Okay, leaving the water piece out for now, who here would be interested in clearing $30,000 rather than $5,000 a year?"

Nobody raised a hand.

Fairly exercised at this point, I asked, "Who here thinks 30,000 is a bigger number than 5,000?"

Nobody raised a hand.

Realizing the dead-end, I withdrew the proposal from consideration, noting that I had asked simply for fair consideration and not gotten it. I was furious.

Looking back now, a decade later, I see my interaction with the John Day agricultural community as indicative of a larger problem. We are part of a system that is physically integrated, yet we treat the parts as if they exist separately. Despite living in the ultimate closed loop (the biosphere), humans operate under an open-ended accounting system; we do not track our impacts well, and we have never accepted in any meaningful way that our natural resources have limits. Culturally, we do not believe that using less could ever equate to prospering more. The farmers and ranchers in that room could not even begin to consider that coupling less water with innovative management could net them more money. For them, more was always better, regardless of where our collective account balance sits.

Over time in a world of limits, such a model will not work. And if we are to tackle the complex water problems that face our planet, the agricultural industry must take a seat at the table. Farmers and ranchers can wait for the situation to become a full-blown crisis. Or they can make the necessary changes now.

For a crystal ball into their future, farmers and ranchers need look no further than the commercial fishing industry. Over the last 50 years, a whopping 90 percent of large fish such as tuna, halibut, and cod have been fished out of the world's oceans, resulting in an enormous economic hit to the industry.[29] Essentially, the fishing industry continued to use more and more technology to systematically harvest more fish than various species could withstand, and the collapses began.

As fish have gradually gone extinct, many in the commercial fishing industry have come to realize that a healthy fishing industry depends on a healthy environment, and they've teamed up with scientists to determine the total catch amount that will still allow for good reproduction. They've also joined with environmental groups to ensure that enough water remains in our streams and rivers.

Unfortunately, it took a crisis for that to happen, but better late to the table than never. Some commercial fishers now say the relationship between a vibrant environment and a vibrant fishing industry is obvious. Said Zeke Grader, executive director of the Pacific Coast Federation of Fishermen's Associations, the most active trade organization of commercial fisherman on the West Coast,

> The fishing industry relies on healthy ecosystems for abundant fish stocks. It relies on sound research and regulations to assure sustainably managed fisheries. . . . And, it relies on clean waters to assure the fish harvested are marketable. All this should be a "no-brainer" for anyone in this industry. . . . The "green" that is driving us, after all, are the greenbacks derived from healthy fisheries—and that is the way it should be, for ecosystem protection only makes economic sense.[30]

Adopting a Quantified Approach

Economic sense, indeed! As is true for the fishing industry, a strong environment is in the agriculture industry's own best interest—it underpins everything. So how do we foster both a sustainable environment and a sustainable agriculture industry? As with modern environmentalism, the first step for today's agriculture is to *assess the current situation*. What was the situation when the Green Revolution began, and what is the situation now? In the years after World War II, both water and energy seemed limitless. New chemical fertilizers made from natural gas and pesticides made from petroleum were just coming onto the market, as was more mechanized farm machinery propelled by cheap fossil fuels. The time was ripe for agricultural productivity to expand.

Yet today, the era of cheap fuel and seemingly limitless water is drawing to a close. We now know the devastating effects that agrochemicals have on our watersheds. And the return on our agricultural investment is starting to diminish. According to research led by global ecologist Jonathan A. Foley and published in *Scientific American*, the average global crop yield over the last 2 decades increased by just 20 percent. "That improvement is significant, but it is nowhere near enough to double food production by midcentury," Foley wrote. "Whereas yields of some crops improved substantially, others saw little gain and a few even declined."[31]

Moving forward, it's unlikely that the rapid gains in crop yields can be sustained. There is only so much more the agriculture industry can do to streamline production using the same old methods. And at some point, we'll no longer have the cheap energy or the freshwater supplies to keep the expansion going.

Many hold out hope that genetically modified (GM) food will drive the next Green Revolution. Work is already under way to develop drought-tolerant and flood-tolerant crop varieties by inserting genes from one species into another. Yet these products are just hitting the market, and it remains to be seen how they will affect the use of water.

Proponents of GM food also claim that GM seeds will reduce the total amount of agrochemicals used, but so far these predictions haven't been borne out. As respected environmental activist Dr. David Suzuki has pointed out, GM seeds are being engineered to be resistant to the herbicides and pesticides sold by specific agribusinesses so that their use necessitates the application of these same companies' agrochemicals. "More than half of the products that are [produced by] Monsanto are seeds that are generated not to produce more nourishment or to be better tasting, but to allow these plants to be drenched with Monsanto pesticides," Suzuki said in an interview with the Canadian Broadcasting Corporation.[32]

Indeed, according to a recent study by Washington State University researcher Charles M. Benbrook, the first 16 years of genetically engineered crops have led to an overall increase in the use of herbicides and pesticides. Overall, between 1996 and 2011, herbicide-resistant crop technology led to a 527-million-pound increase in herbicide use in the United States. Similarly, pesticide use increased by an estimated 404 million pounds, or about 7 percent. "Contrary to often-repeated claims that today's genetically-engineered crops have, and are reducing pesticide use, the spread of glyphosate-resistant weeds in herbicide-resistant weed management systems has brought about substantial increases in the number and volume of herbicides applied," Benbrook wrote. The study predicted that new GM corn and soybean seeds coming onto the market could increase herbicide use by another 50 percent.[33]

An increase in herbicide and pesticide use isn't an option at a time when our waterways are already polluted. Nor is an exclusive focus on *maximizing* crop yields at the expense of our environment. As Hans Herren, a World Food Prize laureate and the director of Biovision, a Swiss non-profit, told *National Geographic* magazine, "The choice is clear. We need a farming system that is much more mindful of the landscape and ecological resources. We need to change the paradigm of the green revolution. Heavy-input agriculture has no future—we need something different."[34]

Moving from Maximized to Optimized

So what should be our outcome? The answer lies in achieving the right balance. Rather than simply *maximizing* crop yields, we must instead focus on *optimizing* them, meaning that we must strive for the highest amount of nutrition within the parameters of a healthy ecosystem. If we want to continue producing enough food to feed the world, we need to preserve both the quality and the quantity of clean freshwater into the future. To do that, we need to redesign the relationship between agriculture and the environment. Rather than prioritizing one over the other, we need to optimize both for greatest benefit. And that means moving away from maximizing food output by whatever means possible to a focus on calories and nutrients per gallon of water.[35]

Optimizing as opposed to *maximizing* yields requires new best management practices for the agriculture industry. We can no longer afford to dump large quantities of fertilizers and pesticides on our crops just because it's cheap. Nor can we afford to irrigate our crops with huge amounts of water during the heat of the day just because it happens to be Tuesday. We need to quantify our actions so that we apply only the fertilizers, pesticides, and water that are truly needed, and do so in a way that optimizes both nutrition and profits while causing the least amount of damage to the environment.

Using Innovation, Data, and Analytics to Farm More Precisely

The good news is that we have the ability to take an optimized approach; it's just a matter of putting this knowledge to use. For more than a decade, for example, researchers have already known how to optimize irrigation for yield and profit. Using mathematical models, they can determine just how much irrigation is needed, taking a variety of factors into account, including both farmer profits and environmental health. As shown in figure 4.1, there comes a point when applying more water to a crop starts to have diminishing returns. Although a rational actor should stop, most agricultural producers keep pouring the water on, partly out of habit, partly in order to maintain an active water right under the law, and partly because they don't track it. If we use detailed models mapping the relationship between applied water, crop production, and irrigation efficiency, it's clear that agriculture can optimize crop yields and profit margins while reducing the amount of water used.[36]

In addition to optimized irrigation, other methods to improve irrigation efficiency abound. For example, drip irrigation, in which water is applied directly to the base of the plant, has been shown to save as much as 80 percent of water compared with other irrigation systems.[37] And the lining of canals and reservoirs can reduce as much as 75 percent of water from seeping into the soil.[38]

Even more efficient yet are the technologies of hydroponics and aeroponics, which can grow plants without farmland or soil by creating a fine mist that delivers nutrients to plant roots. Some are even experimenting with these technologies using multifloor vertical buildings in urban settings, claiming that they use no agrochemicals and as much as 90 percent less water compared with conventional cultivation techniques.[39]

With today's innovations, there's no reason why wasteful irrigation methods such as flood irrigation—in which water is piped to the fields and allowed to flow along the ground among the crops—is still being used. Even though about half of all water used in flood irrigation never makes it to the crops, flood irrigation remains one of the most popular

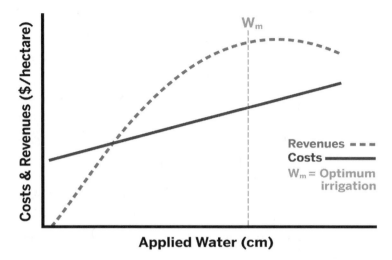

Figure 4.1. Watering beyond optimum irrigation levels wastes both water and money. Though legal, it's not smart. *(Credit: The Freshwater Trust, adapted from Marshall J. English, Kenneth J. Solomon, and Glenn J. Hoffman, "A Paradigm Shift in Irrigation Management," Journal of Irrigation and Drainage Engineering 128, no. 5, September/October 2002.)*

methods of irrigation in the United States.[40] And more broadly speaking, at least half of irrigated cropland acreage across the United States is still irrigated with less efficient, traditional irrigation application systems.[41]

Similarly, when it comes to reducing agrochemical runoff, a wide variety of solutions are available. No-till and reduced-till techniques can help farmers improve soil while managing weeds without the use of herbicides. Buffers of shrubs or trees can be planted between crops and waterways to filter fertilizers and prevent them from seeping into streams and rivers. And cover crops can be planted at the end of the season to prevent agrochemicals from being washed into watersheds by winter rains. As Foley has pointed out, "When nearly half the fertilizer we apply runs off rather than nourishes crops, we clearly can do better."[42]

Another key solution is *precision agriculture*.[43] Similar in some ways to lean manufacturing, which Toyota and others in the business world have implemented to identify and eliminate manufacturing waste, precision agriculture uses the latest technology to use only as much fertilizer, herbicides, pesticides, seeds, and water as needed. Rather than planting a crop and then evenly distributing water and agrochemicals across the field, precision agriculture strives to apply the right treatment in the right place at the right time. By meticulously identifying and eliminating waste while minimizing damage to the environment, precision agriculture promises to take agriculture to the next level.

Using technology such as the Global Positioning System (GPS) and electronic sensors, farmers can measure the soil types, moisture levels, and fertilizer needs in different parts of their fields. They can determine where in the field they're experiencing insect problems. And they can analyze which crops in their field are under stress and why. By linking these sensors to digital mapping programs mounted on their tractors, they can then automatically deliver the exact amount of pesticides, herbicides, seeds, and water that are needed. With such precise inputs, farmers can lighten their impact on the land while reducing their costs.

By reducing their agrochemical use, farmers can limit the chemicals that run off into surface and groundwater. And by irrigating more precisely, they can improve their crop yields while reducing water usage. For example, farmers can use remote sensors to measure how much water their crop is actually using, giving them a more accurate picture of how much more water it needs.[44]

Among those embracing precision farming to conserve water and improve crop yields is Tom Rogers, an almond grower in Madera County, California. Rogers and his brother are using a combination of careful soil moisture monitoring and weather information from on-site stations to help them decide when and how much to irrigate. Readings from soil probes are taken every 15 minutes to provide a detailed picture of how water is moving through the soil. In addition, weather stations mounted in the field provide information on the temperature above and below each tree's canopy, as well as wind speed, humidity, and rainfall, allowing Rogers to keep track of how much water is being added to his fields through precipitation and how much is being lost through evapotranspiration—or the amount of water that evaporates from the soil and transpires from his trees.

All this information is used to determine when to irrigate, saving Rogers up to 20 percent in water use in some fields, with yields higher than many of his neighbors. "We can water according to a calendar, or we can water according to trees' needs," Rogers said. "Our goal here is to water according to the trees' needs."[45]

On the other side of the country, the Iowa Soybean Association has created an On-Farm Network that uses precision agriculture techniques to help soybean growers improve profitability, efficiency, and environmental stewardship. For example, farmers are using satellite images to identify problems in different parts of the field, such as variability in crop growth and damage from excessive rain. Knowing exactly which areas of their field are being lost to water, they can then seed cover crops to prevent weed growth and phosphorus deficiency in those specific areas.[46]

Iowa has also launched a science-based initiative to reduce nitrate and phosphorus loads in the state's waterways by 45 percent, and the On-Farm Network is responding by helping soybean farmers better manage their fertilizer use.[47] For example, soybean farmers are using soil, stalk, and water testing to apply only as much nitrogen as needed to optimize yields. They are installing bioreactors that act like wetlands to filter out nitrates before they reach the river. They're planting cover crops to suppress weeds and build organic matter. And they're embracing "strip tillage," a tilling method that reduces soil erosion and nitrogen runoff by plowing small strips needed to plant the soybeans rather than the entire field.

Arlo Van Diest, a soybean and corn farmer who's been applying these methods, said strip tillage has saved him money while improving the quality of his soil. "We end up with just an entirely different soil structure," he said. "It takes a while. It takes four or five years. You just feel the difference. Actually in the tractor, doing tillage work, you can just tell that our soil is much more mellow and loose."[48]

Breaking Down the Barriers to Participation

Although most, if not all, of these farming reforms increase farmer profits, one of the chief barriers to their adoption is the upfront cost of purchasing the technology. Environmental groups, governments, and investors have the opportunity to significantly move the needle by making funds available for these conservation practices and then helping farmers measure and hone their results.

In addition, organizations can build new technologies and make them available to farmers. In California, for example, the state Department of Water Resources has developed the California Irrigation Management Information System (CIMIS), a compilation of data designed to help irrigators manage their water resources more efficiently. CIMIS takes weather data from more than 140 computerized weather stations throughout California and uses it to provide data about evapotranspiration to help farmers determine when to irrigate and how much water to apply.[49] Growers using these data in a pilot study reduced their water use by 13 percent while increasing their yields by 8 percent.[50]

Another solution that's been proposed, one that tackles incentives from the consumer standpoint, is to create a sustainable food certification program that rates food products based on how well they deliver nutrition, promote food security, minimize water use, and maintain the health of the environment. Just like the Leadership in Energy and Environmental Design (LEED) program, which awards different levels of certification for green building construction, a sustainable food certification program could create momentum for sustainably grown food by giving consumers far more information about how their food is grown. As Foley put it, "This certification would help us get beyond current food labels such as 'local' and 'organic,' which really do not tell us much about what we are eating. Instead we can look at the whole performance of our food—across nutritional, social and environmental dimensions—and weigh the costs and benefits of different farming approaches."[51]

Although similar programs to label sustainably produced forest products and seafood have yet to take off, I believe that is changing and that the buying habits of the next generation will demand greater transparency about the sustainability of food-growing processes.

Managing the Land Holistically

Another way in which the economy and the environment can be brought together is through holistic management, an approach to agriculture that helps farmers and ranchers better manage agricultural resources for long-term economic, environmental, and social benefits. The concept is the brainchild of Allan Savory, a Rhodesian wildlife biologist who wanted to understand what was causing the desertification of the world's grasslands.[52] Desertification, in which dry land turns to desert, threatens about two thirds of all land on Earth, making it unusable for farming or grazing livestock.[53] Today, arable land loss is estimated at 30 to 35 times the historical rate.[54]

Although desertification is typically attributed to overgrazing by livestock, by studying desertification, Savory concluded that livestock are actually good for the land. During times when there were very large numbers of wild grazing animals, their movement prevented overgrazing, and their trampling ensured good cover of the soil. Savory found that by bunching agricultural livestock into larger herds and better planning their grazing to mimic the patterns of wild grazing animals, we can actually reverse the effects of desertification, turning desert areas back into arable land that can support water and vegetation. Savory's organization, The Savory Institute, has been applying the principles of holistic management to 37 million acres across five continents.[55]

The Freshwater Trust has been working with farmers and ranchers in its home region to apply the principles of holistic management to heal grasslands and restore rivers while helping them increase profits. For example, some landowners have agreed to lease their water rights during the summer in order to increase stream flows for salmon runs. Others have agreed to implement water-saving technologies while keeping the same acreage in production. They've then leased their rights to the excess water, making a profit off of that water while allowing more water to remain in-stream.

One farmer we've been working with is Susan Boyd, owner of a 50-acre organic alfalfa farm in eastern Oregon. Boyd used to harvest her alfalfa three times a year, drawing irrigation water from Little Creek bordering her property, which feeds into Catherine Creek. But she's since agreed to lease some of her water rights to The Freshwater Trust during the summer when Catherine Creek runs dry, killing fish that rely on the watershed for habitat. By the third cutting, her alfalfa is usually "pitiful" anyway, so Boyd was open to leasing back the water. Interestingly, her participation in the program initially drew criticism from neighbors—until she told them it's boosted her earnings. "The farmer comes out ahead; the fish come out ahead," she said. "It's a wonderful program."

In addition to working with existing landowners, The Freshwater Trust has also started an experimental program designed to encourage young people who embrace the principles of holistic management to get into the business. The effort is designed to make it affordable for young farmers and ranchers to purchase land by leasing some of their water rights to leave water in-stream.

Programs such as these are aimed at motivating existing farmers and ranchers to change their land management practices while encouraging future generations to implement the right methods from the start—all while turning a profit. It may not look exactly like traditional agriculture, but it works.

Only We Can Do This

Over the past few decades, an ideological debate has been brewing that pits advocates of organic and local foods against advocates of conventional agriculture. Proponents of organic farming argue that conventional farming unnecessarily damages the environment, whereas supporters of conventional farming claim that the lower crop yields of organic farming will never be able to feed the world.

Clearly, organic farming, with its emphasis on environmental steward-ship, is an important part of the equation. And in recent years, organic crop yields have improved. For example, a recent study published by the University of Minnesota's Institute on the Environment found that, al-though organic yields are on average 25 percent lower than conventional yields, the yield gap is much lower for some types of crops, such as soy-beans and fruit.[56]

Yet in my opinion, it's not an either–or situation. If we are to massively increase our food production over the next few decades while improving the health of the environment, the best farming methods from all prac-tices must be implemented. In many cases, conventional farmers can ben-efit by adopting the environmental protection techniques used by organic farmers. Likewise, organic farmers can look to some of the principles of precision agriculture to improve their yields while continuing to care for the soil and water. As the University of Minnesota researchers concluded in their yield comparison study, "To achieve sustainable food security we will likely need many different techniques—including organic, conven-tional, and possible 'hybrid' systems—to produce more food at affordable prices, ensure livelihoods to farmers, and reduce the environmental costs of agriculture."

Not only must we incorporate the best farming techniques across the board, but *all* of us must play a role in developing new solutions. With the worldwide market for agrochemicals expected to grow to more than $240 billion by 2018,[57] we can't afford to shun agribusinesses such as Mon-santo, Syngenta, and Bayer. Love them or hate them, the reality is these corporations aren't going out of business anytime soon. So if it's results we're after, a better solution would be to work with them to change their practices. These companies have huge research arms. By proactively en-gaging them, we have a better shot at developing fertilizers and pesticides that have fewer negative effects on the natural environment. At the same time, we can create a market for these products by providing incentives for farmers to use them.

A great example of this kind of collaboration is Field to Market, a sustainable agriculture alliance of which The Freshwater Trust is a member. The alliance has attracted a wide range of players, including agribusinesses, universities, conservation groups, and others, to help farmers improve the environment while increasing productivity. Among the alliance's innovations is the Fieldprint Calculator, a free tool designed to help farmers better understand the impacts of their farming methods on soil conservation, irrigated water use, water quality, carbon emissions, and other factors.

Field to Market has also launched several projects to help farming be more sustainable. For example, it is working with corn farmers in Iowa's Boone River watershed to establish nutrient management plans that reduce their use of fertilizers. Farmers participating in the project are using the Fieldprint Calculator to understand how their farming practices are affecting water quality. They are also working with Field to Market to implement better management practices, including those used by organic farmers, such as planting cover crops and conservation tillage. Combined, these practices are expected to prevent tens of thousands of pounds of nitrogen pollution from flowing into the rivers.[58]

Attempts to solve our agricultural problems in silos (pun intended) aren't going to succeed at the necessary pace or scale. We need strong partnerships—and many of them—if we're going to feed our booming population without destroying the planet. The ability to manage our portfolio of economic and environmental assets starts with clear quantification of what specifically we get, and what specifically we give up, when we make a decision. New tools and new transactions can lead to outcomes that result in a win–win for agriculture and the environment.

As we will see in the next chapter, other industries are already starting to feel the effects of diminishing freshwater. And, although the business world meticulously quantifies its financial results, it needs to broaden its accounting methods for our economy to continue to thrive in the coming decades.

CHAPTER 5

It's the Environment, Stupid

In the northern Indian state of Uttar Pradesh, the Pollution Control Board recently ordered Coca-Cola to shut down its bottling plant, claiming aquifer levels at the factory were critically low and that Coca-Cola had failed to get permission to use the area's groundwater.[1] The order, which the world's biggest soft drink maker has appealed, followed a similar decision a decade earlier to shut down a Coca-Cola bottling plant in the southern Indian state of Kerala after residents there complained it was draining and polluting local water supplies.[2] In other parts of India, too, Coca-Cola's bottling plants have faced opposition because of rapid water depletion.[3] Amit Srivastava of India Resource Center, which led the campaign to close the Uttar Pradesh bottling plant, explained, "Coca-Cola's business strategy has put its plants near its big markets: the cities. That means it's putting itself in competition with local people and farmers."[4]

Closer to home, another multinational company has also been bumping up against the limits of water. To guarantee its future expansion, Intel, the world's largest semiconductor chip maker, has literally been pumping water back into the aquifer for future use. Rather than reusing all of its wastewater, Intel, with the help of city officials at its manufacturing plant in Chandler, Arizona, has been injecting 1.5 million gallons a day of cleaned-up wastewater 600 feet down into a sandstone aquifer beneath the city. To date, the computer chip manufacturer has saved more than 3.5 billion gallons of water for future use.[5] By banking water for the future, Intel and the City of Chandler are ensuring both Intel's and the city's future growth at a time when population is exploding, droughts are lengthening, and climate change is making water supplies less predictable. "If we never recharge another drop, we have enough water underneath us to last about 100 years," said Dave Siegel, water czar for the City of Chandler.[6]

Examples such as these demonstrate the central, if historically hidden, importance of freshwater to business. Without it, the corporate world would grind to a halt. From automobile manufacturing to consumer packaged goods to high-tech, almost every industry relies on water for at least some part of its process. Water is behind almost every product we buy, yet most people are unaware of it.

Take the $600-billion soft drink industry, for example.[7] One would think that a 16-ounce bottle of pop would take 16 ounces of water for its production. But the reality is that every 16-ounce bottled soft drink we consume takes 45 to 82 gallons to produce.[8] About 95 percent of that water comes from growing the ingredients—mainly the sugar from sugar beets, cane sugar, or high-fructose corn syrup. The remaining 5 percent of water is needed for the production process, labeling, and packaging. So if the 1.9 billion servings per day in 2015 proudly noted on Coke's website were of the 16-ounce cola variety, that would take somewhere between 31 and 57 trillion gallons of water per year to produce. For context, the *low* end of that water footprint range could put the whole state of California under a foot of water. And, although Coca-Cola is the world's best-selling soft drink, it's just one of hundreds of soft drink brands on the market.

Likewise, the needs of the $250-billion semiconductor industry make it an incredibly thirsty business.[9] The microchips that power our computers, laptops, and cell phones rely on large volumes of water of the highest industrial quality. Because even the slightest imperfection can render a chip useless, the water is purified using sophisticated filters that eliminate even microscopic contamination. Large volumes of this water are then used to wash each layer clean as metal is etched away by acid to form the chips' circuitry. Creating an integrated circuit on just one 300-millimeter wafer uses approximately 2,200 gallons of water. For a chip manufacturer such as Intel, that adds up to lots of water. In 2009, for example, Intel reported using more than 7.5 billion gallons of water worldwide.[10]

Nearly every product we buy has a similar water story. Producing a cup of coffee uses 55 gallons of water.[11] A gallon of milk, 880 gallons. A 3-ounce steak, 338 gallons. A pair of jeans, 2,900 gallons. A cotton shirt, 700 gallons. A car, 32,000 gallons. A computer, 39,000 gallons.[12] The electricity a typical person consumes each day, 670 gallons. Every pound of paper, 1,160 gallons.[13]

You get the idea: It all adds up. Simply put, water is the invisible engine that fuels economic growth. The problem is that, as competition for scarce supplies of water continues to mount, the situation could wreak havoc on our global economy. Indeed, a recent report by the Carbon Disclosure Project (CDP), a United Kingdom–based organization that encourages corporations to report on their carbon emissions and water usage, reached a startling conclusion about our global economy's dependence on water. The report predicted that if we continue our current water management practices, an estimated $63 trillion worth of business will be at risk. That's 45 percent of the projected 2050 global gross domestic product (GDP) (at 2000 prices), or 1.5 times the size of today's entire global economy!

After analyzing the water disclosure information of 191 Global 500 companies, the authors concluded that 53 percent have already been harmed by problems related to water, such as business interruption and property damage from flooding, with associated financial costs for some companies as high as $200 million. What's more, 68 percent admit that water poses a substantial risk to their business. "Water-related risks continue to place stress on economies and communities at both local and global scales," the report says. "The financial impacts of floods, droughts, and overall water quantity and quality are rippling across the world."[14]

Outstripping the Supply of Water

The corporate world does an excellent job of maximizing profits—in the short term. Yet businesses have largely failed when it comes to planning for lasting success. Although industry doesn't consume as much water as agriculture, it still has a large impact on both its quantity and its quality. Altogether, U.S. industries use 18,200 million gallons of water per day, or 9 percent of total water withdrawals in the United States.[15] Worldwide, 22 percent of all water is used for industrial purposes. What's more, industrial water usage typically grows as a country's income increases, with industrial water use as high as 59 percent in high-income countries.[16]

The unfortunate reality is that industrial water use isn't expected to slow down anytime soon. The Organisation for Economic Co-operation and Development (OECD) predicts that global water demand will increase by 55 percent by 2050, with some industries needing far more. For example, the amount of water needed for electricity production is expected to increase by 140 percent, and that used by the manufacturing sector is expected to grow by a whopping 400 percent.[17]

This kind of growth cannot be sustained under our current water management regime. In fact, global demand for water is expected to outstrip the supply by 40 percent by 2030, according to a McKinsey & Co. analysis, putting industry at risk as it increasingly competes with communities and farmers for limited water supplies. Said Rose Marcario, CEO of Patagonia, "Business can be the most powerful agent for change, and if business doesn't change, then I think we're all doomed. Business that puts profit above people and the environment is not going to be a healthy and sustainable way for us to live and for the planet to survive."[18]

A Critical Risk Management Strategy

Nobody wants to see a future in which we're all doomed. And we do have a choice here. In the life of any enterprise, there comes a moment when the fundamentals of the operating environment change. Andy Grove, the clear-eyed leader of Intel, called these moments inflection points (figure 5.1). Such moments, Grove admits, are hard to discern from the day-to-day challenges that businesses face. But if not responded to, they can threaten the very life of the enterprise. They force the hard decision to invest in a down cycle and take a more promising trajectory or to do nothing and wither.

Today, the declining availability of water presents a critical inflection point for the economic world. Businesses can adapt to the new reality, or they can ignore it, putting their future livelihoods at risk. To modernize James Carville's famous line from the 1992 presidential election, "It's the environment, stupid."

Making smart decisions about water is critical to corporate risk management. A solid water use plan reduces a company's operational costs. It prevents disruptions to production. It helps companies avoid conflicts

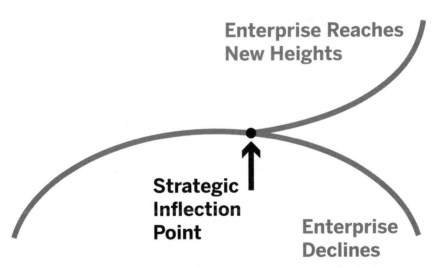

Figure 5.1. Strategic inflection points are moments of fundamental change that must be responded to in order to reach potential. *(Credit: The Freshwater Trust, adapted from Andrew S. Grove,* Only the Paranoid Survive: How to Exploit the Crisis Points That Challenge Every Company, *New York: Doubleday, 1996.)*

with farmers and communities. And it ensures that the company has enough water to secure its operations well into the future. In the long term, the economy will favor businesses that proactively manage their water because they will be the most cost-effective—and the most resilient.

The demand for change is beginning, albeit slowly. Increasingly, investors are requiring that businesses disclose how they manage water. In fact, the number of investors that have called for greater corporate transparency on water has quadrupled in the last 3 years alone, according to the CDP.[19] The strategic question centers not on whether most companies will face water risks but rather on how severe and frequent they will be. Accordingly, for a growing number of investors, the failure to disclose water information will become a reason to divest.

Consumers, too, especially millennials, are beginning to demand sustainability information about the products they purchase. As environmental strategy advisor Andrew Winston wrote in an article in *Harvard Business Review*, "Customers increasingly expect ready access to information about the things they buy. How, where, and by whom are your products made? What's in them? What is their environmental and social impact?"[20] It is increasingly clear this generation not only believes in an ideal but is earnest in bringing it about.

Smart companies that begin responding to these demands now have the opportunity to increase brand loyalty and deepen customer trust, giving them a competitive advantage. At the other end of the spectrum, those that fail to adapt could lose customers. Consider, for example, General Mills's recent decision to stop using genetically modified organism (GMO) ingredients in its original Cheerios. The decision came after 40,000 consumers posted on Facebook in support of a campaign orchestrated by the nonprofit Green America to pressure General Mills to make Cheerios GMO-free.[21] In a world of scarce natural resources, I can see consumers launching these same kinds of campaigns to demand more sustainable products.

Trapped by a Short-Termism

So what's preventing businesses from taking a long-term view of water and other natural resources? In a nutshell, "short-termism." The pressure to make as much money as quickly as possible has discouraged many business leaders from confronting environmental problems that threaten their long-term existence, not to mention success. Preoccupied by short-term performance measures, quarterly earnings, and the demand for immediate profits, today's business institutions are poorly equipped to address the twenty-first-century issues that demand their attention. According to a recent *McKinsey Quarterly* survey of more than 1,000 board members and C-suite executives, 63 percent of respondents said the pressure to generate strong short-term results had increased over the past 5 years, and 79 percent felt pressured to demonstrate strong financial performance over a period of 2 years or less.[22]

A short-term focus worked pretty well in the past. But in a world of limited resources, it's economic suicide. Without long-term planning, an increasing number of businesses may find themselves in the same position as Coca-Cola, in which they compete with farmers and municipalities for dwindling supplies of water. Moreover, they'll have no way of knowing whether they have enough water to expand their operations in the future, let alone sustain their current business.

Broadening the Focus

Today's businesses are at an inflection point, and adapting to the new reality requires a broader focus, one that also accounts for the long-term health of society and the environment. British sustainable development leader John Elkington, who in 1994 coined the term *triple bottom line*, has spent the last couple of decades urging companies to move to a fuller accounting framework that measures their impact on people and the planet—in addition to profits (figure 5.2). Lasting success, he argues, will depend on it.[23]

Figure 5.2. Companies are starting to move to a fuller accounting framework that tracks their social and environmental progress in addition to their financial gains. *(Credit: Image provided by B Lab.)*

Many have since latched on to Elkington's idea, with some corporations now describing their triple bottom line progress on their websites and in their reports. Taking this idea a step further, in recent years the nonprofit B Lab has begun certifying for-profit companies as B Corporations (the *B* stands for the benefit companies provide to society). Cofounded by Jay Coen Gilbert, Bart Houlahan, and Andrew Kassoy, B Lab is working to redefine business success by marrying "the power of markets with the purpose and mission of the nonprofit sector."[24] To earn the certification, companies must fill out a lengthy questionnaire that covers everything from charitable giving to energy efficiency to employee benefits in a process similar to the Leadership in Energy and Environmental Design (LEED) certification system for green buildings. They must also amend their articles of incorporation to say that managers must consider the interests of employees, the community, and the environment rather than focusing exclusively on shareholders. So far, more than 1,200 companies have received the certification, including popular brands such as Patagonia, Etsy, and Seventh Generation.[25] As more businesses become certified, B Lab's founders hope it will give consumers a better way to make buying decisions. "We want to help consumers separate good companies from good marketing," said Coen Gilbert.[26]

In addition to certifying businesses, B Lab has been hard at work changing state corporate governance laws. In some states, corporate boards of directors are required by law to maximize shareholder value over all else, limiting their ability to create benefits for society if it means accepting a lower financial return.[27] B Lab has been working to change that by convincing states to allow companies to incorporate themselves as "benefit corporations." Large private equity, venture, and institutional investors are beginning to invest in these corporations, understanding the commitment they make to their stakeholders is their value. In just 4 years, 30 states have changed their corporation forms to allow benefit corporations.[28] With these changes in place, shareholders in those states will be able to sue their corporate boards of directors if they fail to carry out their social mission in the same way that they can sue traditional companies for failure to perform their fiduciary duties.[29] Now that's corporate social responsibility with teeth!

Tracking Environmental Health

As we broaden our definition of success, we must also find a way to quantify it. History has shown that, when the value of a resource isn't measured, we don't invest in protecting it. In fact, we aggressively consume it until it abruptly becomes scarce.[30] Accordingly, any meaningful environmental reform requires measurement and quantification. To bring about greater environmental and societal health, we need indicators that help us gauge our success at both the macro and the micro levels.

On a national scale, this means measuring and reporting on our environmental progress alongside indicators such as GDP and the consumer price index (CPI) that countries rely on to assess their financial prosperity. Indeed, some organizations have been hard at work developing such measures. For example, Yale and Columbia universities jointly publish an annual environmental health index (EHI) that ranks how well countries protect ecosystems and human health from environmental harm.[31] In addition, the OECD teams up with other international organizations to publish environmental indicators in areas such as water consumption, fisheries, and carbon dioxide emissions.[32] For these indicators to have relevance, they must be given the same prominence as GDP and CPI, and nations must disclose their progress on a regular basis.

At the company level, accountability means reporting on environmental metrics as well as on financial results. Just as businesses measure their financial success through quarterly income, cash flow, and stockholders' equity statements, they need to continually monitor their supply of clean freshwater, their carbon emissions, and their electricity use. Here, too, numerous indicators have cropped up over the past several years. There's the CDP, the Global Reporting Initiative, the Dow Jones Sustainability Index (DJSI), the UN Global Compact, the Alliance for Water Stewardship, and the International Organization for Standardization, just to name a few.

Many companies are starting to use these indicators to report on their progress, which is an important first step. And global accounting firms have been taking these practices mainstream with accounting methods such as PwC's "total impact measurement."[33] Yet, without a common definition of water stewardship, businesses are reporting their results in different ways. As a result, the depth and focus of business reporting vary, and it's not uncommon for businesses in the same industry to use different measurements in their reports, making it difficult to compare performance.[34]

In the same way that we have widely accepted international standards for product safety and quality, we need a single set of standards to measure businesses' environmental impact. If all businesses adhered to the same standards, for example, lending institutions could easily rank, score, and compare them in a way that informs their investment decisions. Similarly, consumers could choose to work with companies with the best track records.

Reexamining the Business Relationship to Water

Without a single set of universal standards and parameters at the national or corporate levels, true natural resource accountability has yet to take hold. In the meantime, the majority of today's businesses continue to ignore today's environmental realities, to both their own peril and the planet's. Businesses' attitude toward the environment reminds me of a bad relationship. When it comes to our personal lives, most of us realize that a good relationship is a matter of give and take. Healthy relationships involve compromise, and they require mutual respect. Yet the relationship the economy has with the environment has been all take and no give. It really amounts to an abusive relationship, and the economy remains oblivious. It's as if the corporate world doesn't even understand that it's *in* a relationship. As a result, the environment is forever trampled upon, without anyone saying, "Enough already."

Although the inertia of status quo persists, the system is inching toward change as a few forward-leaning companies take steps to beef up their environmental stewardship. These early adopters are still by far the exception rather than the rule, which means that even in the aggregate, they're not yet moving the needle. Nevertheless, their efforts are slowly charting a new path and are worth calling out. In my opinion, companies that are the most serious about water stewardship are doing three things: They're assigning a value to water, shifting it to a CFO issue rather than simple PR greenwashing. They're taking a comprehensive, integrated approach. And they're getting buy-in at the board level.

Putting a Price on Water

Today's markets don't factor in the value of water either to businesses or to their surrounding communities and ecosystems. As a result, water is basically seen as a free raw material. Yet as water becomes scarce, corporations are beginning to encounter resistance from the communities in which they operate, and governments are starting to set limits on its use. Companies that quantify their dependence on water can obtain a more complete picture of their business risk. They can also make more informed decisions, such as where to locate facilities and what water-efficient technologies to use in their production processes. Simply put, water has an economic value, and we should be tracking it.

Because the market doesn't assign water a value, some companies are using "shadow pricing," in which they estimate the value themselves. By putting a price on water, businesses can easily calculate and present the costs and benefits of water in their operations alongside other business metrics, giving them an easier way to make decisions. For example, Nestlé has set the price of water at $1 per cubic meter in facilities where water is abundant and $5 where water is scarce. Assigning a value to water has allowed the company to better evaluate proposals to buy new equipment that improves water efficiency in its factories.[35]

Likewise, the United Kingdom–based utility Yorkshire Water prepares environmental profit and loss statements to help it gauge its water impact. The company records negative environmental impacts, such as water extraction and pollution impacts, as losses, while on the profit side of the ledger are environmental benefits, including water recharge and operational energy recovery techniques. By putting a monetary value on water, Yorkshire Water can better gauge the water impacts of its supply chain. At the same time, the company has found a more meaningful way to communicate its strategy to its suppliers, customers, and regulators.[36]

A Comprehensive Approach

Companies serious about water stewardship are also examining their impact on water from a wide range of perspectives. They're evaluating water use not just in their direct operations but all along their value chains, from their raw materials to their finished products. They're also helping communities improve the health of rivers in the regions where they do business.

To help companies develop robust water management plans, the United

Nations in 2007 launched the CEO Water Mandate, a public–private initiative that has identified six key ways in which businesses should improve their water management: reducing water use in their direct operations, encouraging suppliers to improve their water monitoring and conservation practices, working with nonprofits and governments to take collective action, helping to shape strong public policies, partnering with local communities to improve watersheds, and encouraging transparency by publishing and sharing their own water strategies.[37] As companies draw up their roadmaps, many are looking to these six areas as a framework for charting their progress.

Board-Level Buy-In

The most earnest businesses are also creating accountability at the highest levels of the company. As executives at Nike have learned firsthand, board-level participation in sustainability efforts can go a long way toward driving greater accountability and innovation. By creating a corporate responsibility committee with a direct line to the board, Nike executives have kept sustainability near the top of the company's list of priorities.

As an example, Nike's corporate responsibility committee persuaded the board to invest in DyeCoo Textile Systems, a startup that developed a waterless process for dyeing polyester using recycled carbon dioxide. With prompting from the corporate responsibility committee, Nike made this decision even though DyeCoo's dyeing process wasn't cost-competitive at the time. At least from a water perspective, the investment is paying off. DyeCoo's dyeing process saves 8 to 12 gallons of water per pound of fabric and eliminates chemical discharges from entering the water supply. As a result of the investment, Nike could save billions of gallons of polluted water from entering watersheds near its manufacturing plants.[38] "By asking insightful questions, making suggestions, offering perspectives, raising counterpoints, and proposing alternatives, the committee enriches and challenges management's thinking," said Harvard business professor Lynn S. Paine, who studied the work of Nike's corporate responsibility committee.[39]

Reducing Their Water Footprint

Once a company is armed with a way to value water, a solid roadmap, and board-level buy-in, the next question is where to begin. Here again, it starts with situational awareness. To better understand how water underpins the business, companies need to measure its use. In fact, measurement by itself can set the wheels in motion for a brand new relationship with water. As author and *Fast Company* staffer Charles Fishman put it, "Measuring alone creates an imperative for curiosity and innovation, and for changing behavior. Just as when you keep track of every calorie you eat, you start cutting back. Just as when there's a real-time miles-per-gallon number on a car's dashboard, you can't help but drive in such a way as to keep the number high."[40]

Water use can show up in so many places, making its measurement complex. To tackle these issues, forward-thinking companies are strategically evaluating their water use throughout the entire lifecycle of their products to see where their footprint is largest. As the result of these analyses, some companies are cutting back water use in the production process. Others are examining water efficiency in their supply chains. Still others are developing new products that encourage consumers and businesses to minimize their water use. The best companies are holding themselves accountable with ambitious goals and aggressive timelines.

MINIMIZING THE WATER USED IN PRODUCTION

The place where companies are paying the most attention to water is in their production processes. Today, about two thirds of Global 500 corporations have taken steps to lower their use of freshwater in their direct operations.[41] Most have accomplished this by reducing the amount of water needed or by recycling and reusing their water many times over.

One company that's reduced the water used in the manufacturing process is Intel. Since 1998, the company has invested more than $100 million in water conservation programs at manufacturing plants around the globe. For example, Intel recycles about 25 percent of its ultrapure water needed to manufacture its microprocessors, using it for air scrubbers to reduce particulate emissions and in its cooling towers to provide air conditioning for its buildings. It also reclaims recycled gray water from publicly owned sewage treatment plants, using it for these same purposes and for building landscaping. To date, these efforts have saved 36 billion gallons of water—enough water to supply 335,000 U.S. homes for an entire year. The savings represent a 40 percent reduction in the water the company uses in its manufacturing process.[42]

Another company that's reduced water in its operations is Ford Motor Company. After assessing its water footprint throughout the life cycle of its vehicles, Ford set the goal of reducing the water used in its manufacturing process by 30 percent from 2009 to 2015, or 1,056 gallons per vehicle. The company is accomplishing this by lowering the amount of water used in everything from cooling towers to parts washing to paint operations. For instance, Ford has implemented a process called minimum quantity lubricant, which lubricates cutting tools with a fine spray of oil exactly when and where it's needed. This alone has cut water use by 282,000 gallons per year for a typical production line of 450,000 vehicles. The company has also consolidated painting activities in one booth, reducing the number of booths that need to be washed out. Thanks to practices such as these, Ford is on its way to reducing the water used per vehicle by about one third by 2015.[43]

EXAMINING THEIR SUPPLY CHAINS

Often, the biggest water saving isn't in a company's own operations but in the raw materials a company buys to make its product. Accordingly, some companies are requiring their suppliers to report their water use. In some cases, businesses are working with their suppliers to lower their water dependency. In other cases, they're setting strict requirements that their suppliers must follow.

Many businesses at the forefront of working with their supply chains are beverage companies. For example, after measuring water usage across its value chain, the beer company MillerCoors discovered that more than

90 percent of water use was occurring in its agricultural supply chain. To address this issue, MillerCoors has since partnered with The Nature Conservancy to determine whether it could reduce water used to grow barley in Idaho's Silver Creek Valley, where most of the company's barley is grown. MillerCoors also developed a Showcase Barley Farm to serve as a model of water conservation practices for interested farmers. By implementing precision agriculture irrigation methods and installing Global Positioning System (GPS) technology that can be used to remotely manage irrigation systems, the Showcase Barley Farm has saved a total of 429 million gallons of water.[44]

These same practices have been implemented by other barley farms that supply MillerCoors. Ideally, brewers should be requiring all of their producers to meet these standards. The cost of upgrades could be defrayed by adding another penny a bushel to the standing contract—a small price to pay to ensure their producers are squeezing the most out of every drop. For businesses, the key driver will always be economic self-interest. Shoring up their supply chains is the goal, and keeping water in-stream is a bonus.

DEVELOPING PRODUCTS THAT USE LESS WATER

Water doesn't only go into the production process; finished products themselves also use water once in the homes of consumers. To address this problem, some companies are working to develop products that encourage consumers to use less water—from water-efficient dishwashers, washing machines, and toilets to laundry soap and personal care products.

One such company is the consumer goods corporation Unilever. By measuring water use across its value chain, Unilever determined that the largest portion of its footprint comes from consumers who shower, bathe, and wash clothes using its products. Accordingly, it has set the ambitious goal of halving all water associated with the consumer use of its products by 2020. To accomplish that goal, the company is developing laundry detergents that require less rinse water. It's also trying to reduce consumer water use by introducing products such as dry shampoo and foam hand-wash, which can cut water use by as much as 18 percent.

Unilever has also taken steps to cut back water in both its supply chain and its manufacturing process, especially in water-scarce countries. For example, the company is working with its tomato suppliers to replace traditional watering practices with drip irrigation. It has also reduced the amount of freshwater extracted by its manufacturing sites by 73 percent over 1995 levels, mainly by treating and then recycling water. By shrinking the amount of water that goes into the manufacturing process, Unilever, since 2008, has saved nearly half a gallon of water for every person on the planet.[45]

This isn't just trendy, green consciousness. The efforts have actually saved Unilever money—to date, about $21 million. Even more important, they're preparing the company for the future. "If you strip everything back and ask yourself, 'Why is Unilever doing this?' It is doing it to prepare itself for a world that will be very different from the world we live in," said Gavin Neath, senior vice president of sustainability at Unilever. Neath foresees a world in which the food, energy, and water that we take for granted today will be in very scarce supply: "We are preparing ourselves for that future. We're also trying to develop products and services which will allow our consumers to adapt to a very different world."[46]

In addition to consumer products, the market is growing for products that help businesses adapt to a water-scarce world—from water management software to water desalination and wastewater treatment technologies. For example, one breakthrough innovation has sprung up at San-Francisco–based WaterFX. WaterFX has created a solar-powered desalination system that treats drainage water from farms. The system, which produces nearly 500 gallons of clean water per hour, overcomes one of the key obstacles to water desalination: the high cost of fuel. Aaron Mandell, founder of WaterFX, said he began working on the technology after concluding that water is the most significant limiting factor to economic growth: "Water will always use energy, so the question is how do we sustainably deal with the water we use—it comes down to consumption and linkage between water and fuel."[47]

Investing beyond the Business

As the CEO Water Mandate and several other water stewardship initiatives have made clear, a robust water strategy involves more than minimizing a company's own water footprint. It also means taking an active role in protecting rivers and streams. It means forming partnerships with governments, nonprofits, and other businesses to share best practices and technologies. And it requires advocating for effective water policies at all levels of government.

IMPROVING WATER AT THE BASIN LEVEL

Several organizations that work on water issues have urged businesses to get to know their river basins, including the health of these rivers and their impact on them. Not only does watershed restoration contribute to the health of communities, but it ensures that businesses have enough clean freshwater to sustain their operations into the future.

Although most businesses continue to focus on minimizing water in their direct operations, a few pioneering companies have responded to these calls by investing in the protection of local watersheds. For example, Coca-Cola has set the ambitious goal of returning as much clean water as it uses back to the environment by 2020. To date, Coca-Cola has returned 108.5 billion gallons of water back to local watersheds—about 68 percent of its total sales volume.[48]

As part of these efforts, the soft drink maker has been partnering with governments and nonprofits to conduct water vulnerability assessments, implement water protection plans, and participate in river conservation work.[49] For instance, Coca-Cola has been working with the World Wildlife Federation to inspire local farmers in the Mesoamerican Reef, which stretches 700 miles from the tip of Mexico's Yucatán Peninsula to the coastal waters off Guatemala and Honduras, to reduce their water and agrochemical use. Said Greg Koch, Coca-Cola's director of global water stewardship, "Our entire value chain . . . is dependent on water. From the product to the manufacturing, the ingredients, you name it. We invest and partner on . . . watershed issues because they're vital for the health and growth of our business and the communities we are a part of."[50]

ESTABLISHING A LEADERSHIP POSITION

Beyond the four walls of their companies and their value chains, and even beyond the watersheds vital to their success, corporations can and should play a leadership role. When it comes to solving the world's water problems, the corporate world has a lot to bring to the table, including strategic planning, an innovative mindset, and considerable financial power. All of this must be placed into action if we're to address our water issues at the pace and scale needed for success.

A great model is the Risky Business Project, a group of business leaders and former policymakers ranging from Cargill executive chairman Greg Page to former U.S. Treasury secretary Hank Paulson working to prepare businesses for climate change. Although the participants don't agree on all the solutions, they're working together to raise awareness of the problem, and their high-profile leadership has been focusing national attention on this critical issue.[51]

These same kinds of partnerships are needed to better manage our water. For example, businesses can encourage governments to take action by lobbying for high-quality water reporting standards. They can participate in the development of well-designed water trading programs that protect both the quantity and the quality of freshwater. And they can develop innovative technologies that bring about a world in which water is managed far more precisely and thoughtfully than it is today.

Although they are small in number, a few businesses are addressing these challenges by joining international working groups, forming industry alliances, and advocating for stronger water policies. For example, Coca-Cola, MillerCoors, Ocean Spray, and a host of other companies are participating in an alliance aimed at advancing environmental sustainability in the beverage sector. The Beverage Industry Environmental

Roundtable (BIER) was formed to influence environmental sustainability standards while sharing best practices that encourage water and energy conservation, sustainable agriculture, recycling, and other sustainability matters that affect the industry.[52] Among the solutions BIER is working on include establishing consistent methods for measuring water footprints across the beverage industry. It has also developed a process to help the beverage industry seize water-related opportunities while managing water risks.[53]

Likewise, Nestlé is establishing a presence at high-profile water initiatives, such as the CEO Water Mandate, where it is helping to draft business best practices that guarantee the human right to water and sanitation. It also recently helped the International Organization of Standardization to craft new specifications for how businesses assess and report their water footprint.[54]

Nestlé has taken on this leadership role, at least in part because its chairman believes water is an urgent yet unrecognized crisis. As Nestlé chairman Peter Brabeck recently told *The Financial Times*, "We have a water crisis because we make wrong water-management decisions. Climate change will further affect the water situation but even if the climate wouldn't change, we have a water problem and this water problem is much more urgent."[55]

A Seismic Shift in Sustainability

Although the world holds many examples of companies working to reduce their water footprint and make their supply chains secure and sustainable, Brabeck is right about the lack of urgency. We're not moving far enough fast enough. A mere 4 percent of Global 500 companies have set goals for their supply chains and 2 percent for watershed management.[56] What's more, less than a third of the largest publicly traded companies in the United States formally oversee sustainability performance at the board level.[57] To me, it seems obvious that they should, if for no other reason than to ensure that their supply chains are secure and that their footprints do not expose them to legal or public relations hassles further down the road.

Still, statistics such as these prompt the question, Is the corporate world really serious? Although companies are steadily building their sustainability plans, in most cases these efforts aren't keeping up with the present predicament. In a world of permanent resource scarcity, incremental changes won't be enough. As Andrew Winston, who's spent more than a decade exploring how large companies deal with environmental and social pressures, put it, "An extreme world calls for extreme change."

If you categorize extreme change as shifts by a factor of 10, we need to move beyond a few one-time examples to broad adoption, and quickly. To bring about a world in which every business manages its water, such efforts must become a requirement. Consumers can play a role through the products they buy. Retailers and manufacturers can require it of their suppliers. Most importantly, governments and investors can make it mandatory.

Intuitively, we all sense it: Doing right by the environment is the only viable long-term strategy. The difficulty is that no generation before us has had to intentionally and collectively close the accounting loops. Did James Dean think about how many gallons of water went into his first pair of jeans? Did Ray Kroc consider the impact of beef patties on the environment when he tasted his first burger? I know I didn't. Our system just isn't set up like that. Our options until now have been molded by a framework built by earlier generations to deal with the immediate needs of *their* times. Largely invisible to us, the defaults were set up to bring about an economic, environmental, and cultural undoing—not just in the United States but across the planet. What we need is a new framework that regards our biosphere as our budget and accounts for inputs, impacts, and outcomes. By resetting the defaults, we can recalibrate our actions to generate a new and more durable wealth in this century—while giving us an honest shot at prosperity in the centuries that follow.

History has demanded many things of us, but the demands going forward are of a fundamentally new kind, and they are coming at us with brutal ferocity. My dad used to say that if you don't change your direction, you'll end up where you're headed. True enough. And unless we want to end up where we're headed now, we need to get much better at understanding where our actions have us pointed and what we need do to correct course. The bottom line is that businesses cannot succeed in a world that's failing.

CHAPTER 6

Throwing Money at the Problem (and Missing)

ONLY ONE DISEASE in human history—smallpox—has ever been completely eliminated. So eradicating several diseases at the same time is an incredibly ambitious goal. So ambitious, in fact, that no one has ever taken it on.

No one, that is, until Bill Gates.

Gates, who started the Bill and Melinda Gates Foundation with his wife in 2000, has outlined a series of impressive goals. He's working to dramatically reduce the spread of HIV and tuberculosis. He's set the target of wiping out polio by 2018. And he wants to eradicate malaria within his lifetime. "Zero is a magic number," he told the *Financial Times*. "You either do what it takes to get to zero and you're glad you did it; or you get close, give up and it goes back to where it was before, in which case you wasted all that credibility, activity, money that could have been applied to other things."[1]

Gates has never been one to squander his money or his effort. And with his foundation, he's applying the same rigor that sustained Microsoft to improve the future for the world's poorest children.

One way Gates is finding success is through innovation. For example, even when vaccines are available, a major obstacle is keeping them cool enough to prevent them from spoiling in areas of the world without electricity. To address this issue, the Gates Foundation partnered with the invention company Intellectual Ventures to build a vaccine storage device that keeps vaccines fresh for more than a month at a time.[2] Another challenge to the spread of disease is the lack of safe, affordable sanitation. To address that issue, the Gates Foundation launched a contest, awarding grants to six researchers who designed low-cost toilets that operate without plumbing.[3]

Another tool that Gates brings to his foundation is scale. In the same way that Microsoft tests its products before pouring money into them, the Gates Foundation looks for practices that are proven to work and then funds them on a massive scale. For example, the foundation realized that more people turned out to its polio vaccination drives in India when local mobilizers were sent into neighborhoods to rally support for them. With that information in hand, the Gates Foundation extended the practice across India, boosting the number of children who got the vaccines.[4]

Yet another tool in the Gates Foundation toolbox is data and analysis. Vaccinations are effective only if children get them. To measure its success, the foundation uses statistical sampling to determine whether adequate coverage levels have been reached. Likewise, the foundation uses measurement to understand where in the vaccination process costs are highest so it can improve future efficiency.

Before the Gates Foundation got involved, global efforts to combat these diseases were reaching a dead end. Yet today, the situation is gradually turning around. Polio has been completely eradicated in India, with efforts continuing in Afghanistan, Nigeria, and Pakistan.[5] Similarly, malaria deaths have decreased 42 percent over the 12-year period from 2000 to 2012.[6] If current trends continue, Gates may just reach his goal of eradicating malaria within his lifetime. "I want to admit that I am an optimist," Gates told *CBS News*. "Any tough problem—I think it can be solved."[7]

Just think if that same rigor were applied to protecting our streams and rivers. What if every investment, whether public or private, were tightly focused on an ambitious goal? What if every dollar spent were measured against that goal? And what if the initiatives that held the most promise were quickly replicated on a broader scale?

Neither public nor private funding works that way today. For the most part, funding is disconnected from performance, and the success of grants isn't adequately measured. Without thorough evaluations, there's no way to inform future giving. Nor is there any way to decide which projects ought to be replicated.

It wasn't always that way. Like Gates, some of the early philanthropists were entrepreneurs who applied both their business acumen and their wealth to social problems. For example, steel tycoon Andrew Carnegie, who believed in giving to the industrious and the ambitious, set his sights on building public libraries. Over a 20-year period, nearly 1,700 libraries were built with his support.[8] Likewise, John D. Rockefeller, co-founder of Standard Oil Company, focused part of his foundation's efforts on public health, particularly the fight against hookworm. At the time, 40 percent of people in the southern United States were infected with the parasite, causing everything from digestive problems to stunted growth. Using a three-pronged approach that included mapping the disease, curing patients, and funding education campaigns to stop the parasite's spread, the Rockefeller Sanitary Commission greatly reduced hookworm in the southern United States within 4 years. With its goal accomplished, Rockefeller closed down the commission, replacing it with a new organization that focused on other global health issues.[9]

The early philanthropists were serious about getting results. They thought big. They had clear goals. And when they set out to accomplish a goal, that's what they did.

A Spray and Pray Approach to Philanthropy

Unfortunately, today's public and private funding isn't nearly as ambitious or goal oriented. In the United States, governments, foundations, and individuals spend billions of dollars on environmental causes each year. Yet in many cases these funds aren't allocated strategically, nor are the results of these efforts adequately measured.

Awarding Grants without Results

Take government grants, for example. Rather than allocating grant money strategically, governments often cast a wide net in deciding who can apply for grant money and then fail to adequately monitor how this money is spent. The U.S. Department of Agriculture (USDA) is a case in point. It and its associated agencies spend nearly $6 billion a year in conservation programs designed to improve watersheds and foster sustainable agriculture.[10] Yet they typically dole out this money very broadly—to, say, any farmer whose property borders a river or any rancher who has a natural resource concern.[11] As we've discussed earlier, not all restoration projects are created equal. For example, planting trees on the south side of a river will probably do far more to reduce river water temperatures than planting the same number of trees on the north side. Despite this fact, there's little analysis to determine which restoration projects along a specific river will have the most benefit. And without these data, it's impossible to judge whether taxpayer money is being wisely spent.

Governments also often fail to adequately measure the success of their grant programs. Without rigorous evaluations, it's impossible to tell which programs are worthy of future funding and which ones ought to be canceled. For example, a recent study of the impact of farm bill grants on California water quality found that two USDA programs aimed at combating agrochemical pollution, the Environmental Quality Incentives Program (EQIP) and the Agricultural Water Enhancement Program (AWEP), weren't being allocated effectively. Rather than spending the money on land management practices such as nutrient management, cover cropping, and filter strips, the study found that grant recipients were spending the lion's share of this money on structural installations such as cement infrastructure, irrigation equipment, and animal fences. Even more outrageous, the authors of the study had wanted to evaluate a third grant program, the Conservation Stewardship Program, but the USDA wasn't able to provide detailed information as to how those dollars were being spent![12]

In this day and age, there's no excuse not to measure the effectiveness of government grant programs. With today's analytic tools, governments can easily determine which programs will have the most impact, even before the money is allocated. They can also monitor the effectiveness of government grants over time, making sure taxpayers reap a maximum environmental return on every dollar spent.

Wasting Billions on Ineffective Philanthropy

We as a society rely heavily on foundations to solve our environmental and social problems. As Michael E. Porter and Mark R. Kramer describe in their *Harvard Business Review* article "Philanthropy's New Agenda: Creating Value," foundations are in a unique position to lead social progress. They're bestowed with a lot more money than individual givers. And they're free from the political pressures facing governments, giving them extraordinary independence to explore new solutions to social problems.[13]

Yet, although U.S. foundations donate about $50 billion a year to social causes, few take full advantage of their unique position. Foundation funding is typically short term and fragmented. For one, grants are typically awarded for short durations of time. The vast majority—95 percent—must be reapplied for after 1 year.[14] And the average multiyear grants are only 2.5 years in length.[15] What's more, grant sizes are too small to have maximum impact. The median size grant is $50,000, far too little to solve our watershed problems, let alone any environmental or societal problem, at a scale that matters.[16]

Bombarded with tens of thousands of grant applications, foundations too often split their funding over a large number of small programs across a large number of sectors. The average foundation makes grants in ten unrelated fields each year. Foundations have been accused of taking the "spray and pray" approach—spraying their funding very broadly, and then praying for results. As Harvard professor Michael E. Porter put it,

"The real scandal is how much money is pissed away on activities that have no real impact. Billions are wasted on ineffective philanthropy."[17]

Not only is grant money wasted, but today's grant-making process prevents nonprofits from being as effective as they could be. With the constant need to reapply for multiple, soon-to-expire, short-term grants, nonprofits are locked into an endless cycle of fundraising. In fact, some nonprofit executives report spending more than half of their time on fundraising.[18] Not only is this frustrating, but it's an ineffective way to operate. To achieve greater impact, nonprofits need to be freed from the fundraising treadmill so they can focus on the problems they were hired to solve.

Both foundations and the nonprofits they support would be more successful if foundations chose just a few social problems and then developed strategic ways to solve them. Rather than just praying for the best, they could use data and analysis to inform their results. Yet here, too, traditional foundations fall short. Although it's common for foundations to track the amount of money they spend and the number of projects they complete, few measure the effectiveness of that spending or the environmental or social impact of their grants. And even fewer measure each grant's impact on the foundation's overall goals. In fact, many foundations consider measurement as something unrelated to their charitable missions.[19]

Without measurement, it's impossible to know which grant programs achieved the most social good for the dollar. The lack of evaluation means that foundations are more likely to award grants to nonprofits with whom they already have a relationship as opposed to projects that show demonstrable impact.[20] As the leader of one nonprofit put it, "I've grown incredibly frustrated by the total disconnect between performance and access to capital in the nonprofit world. We double every single year, we get better impact measurements, and still no one ever comes back to us and says, 'Hey, you guys are doing so great, we want to give more. We want to invest more.'"[21]

Indeed, many foundations are willing to fund early-stage programs, especially those that are low-risk. Few set aside funds for higher-risk projects or for helping nonprofits build the organizational capacity they need to scale up their most effective programs.[22] In general, traditional foundations operate conservatively, sticking to the way they've always done things and failing to encourage innovation on a large scale.

When it comes to overall spending, traditional foundations are risk averse too. On average, they spend just 5.5 percent of their total endowment on an annual basis—a half percentage point above the minimum 5 percent that the Internal Revenue Service requires.[23] In some ways, this makes sense. If foundations want to keep on giving over the long term, they need to preserve their endowments. On the other hand, holding so much of their assets on the sidelines does little to create social value.

Rather than investing most of their endowments in traditional financial markets, foundations should consider freeing up more of these funds to support their philanthropic work, for example, by providing low-interest loans to nonprofits. This type of program-related investment (PRI) works well for impact initiatives that bring in revenue and have the potential to scale.[24] Additionally, foundations should review their portfolios to invest in financial instruments that, at a minimum, are not at cross-purposes with their missions and, ideally, look for mission-related investments that generate environmental and social returns of the kind they would like see in the world.

Investing without Visibility

Across the United States, the largest source of giving comes from individuals, to the tune of more than $240 billion a year.[25] Yet, for the most part, individuals don't have much to go on when it comes to assessing the effectiveness of their donations. In most cases, they don't know exactly how their money is spent. And they aren't told what effect their donations end up having on the problem they want to solve.

Some organizations have been set up to evaluate nonprofits. For example, BBB Wise Alliance accredits nonprofits based on their ability to meet twenty accountability standards, including how well they measure their own effectiveness. Charity Navigator rates nonprofits according to their financial health and their accountability and transparency. CharityWatch assigns a grade to charities according to factors such as the percentage of revenue spent on their charitable purpose. Each of these groups use a different set of criteria, which in some cases isn't at all related to the nonprofit's success at achieving its goals.

Without an effective way to evaluate and compare nonprofits, individuals end up making donations based on the wrong criteria. For example, just 11 percent of individual donors say they make donations based on the results a charity gets. Instead, most base their decision on other factors, such as how well known the charity is or how little it spends on overhead.[26] This is somewhat understandable. No one wants to waste their donations on organizations that use their precious money on plush offices and exorbitant staff salaries. Yet some overhead expenditures are necessary to build the organization's capacity—including staff training, program evaluation, and technology—so they can sharpen their impact as they grow. A better way to gauge nonprofit success is by the social return on its investment—in other words, by how much good is accomplished per dollar spent.

A Quantified Approach to Funding

Over the past couple of decades, a growing number of thought leaders have been calling for philanthropy to be reformed, urging philanthropists to follow the lead of business. As Harvard professor Michael E. Porter told the *Economist*, "Philanthropy is decades behind business in applying rigorous thinking to the use of money."[27]

Indeed, new approaches to philanthropy have been cropping up with names like "philanthrocapitalism," "venture philanthropy," and "impact investing." The common thread among these approaches is transforming philanthropy so that it mirrors the way that business is done in the for-profit capitalist world. As part of these reforms, this next wave of philanthropy is moving away from trying to be all things to all people and instead taking a focused approach. As the Gates Foundation has done with its focus on disease, this involves setting ambitious goals and making them measurable. Many new-style philanthropists are moving away from risk-averse grants to funding experimental, new ways of doing things. Some are taking a hands-on approach similar to the relationship that venture capitalists have with the startup companies they fund. Many are taking the most promising projects and replicating them on a larger scale. And almost all are demanding a maximum social return on their investment in the same way that businesses require a financial return.

The Robin Hood Foundation is considered one of the darlings of this new-style philanthropy. Started by billionaire hedge fund manager Paul Tudor Jones, the foundation is focusing its funding on a single challenge: fighting poverty in New York City. With its clear focus, the foundation has found what it believes are the most effective 200 organizations working toward that goal, and it is working closely with them to increase their effectiveness even further so they can assist more people. Robin Hood works with its grantees just like an active hedge fund would—by expanding organizations that are thriving, stepping in to provide advice to ones that aren't, and eventually cutting nonprofits loose if they fail to get results. It recommends new executive directors and even matches business-people with nonprofit boards of directors.

To ensure its success, the foundation hires an outside evaluation firm to collect data about the organizations it supports. Much as businesses calculate their financial returns, Robin Hood compares the value of grants by estimating their benefit to the poor per dollar of cost to the foundation. Specifically, it uses a benefit–cost ratio to compare the dollars it contributes to a program against the potential future earnings of the poor people who participate in it. "We set out to compare the relative return of any two grants, no matter what their purpose, much the way an investor compares the relative return of any two investments," said American film producer Harvey Weinstein, who serves on the board. "For most of our grants, our basic measure of success is, How much does our grant boost early adult earnings of poor individuals?"[28]

To date, the foundation's accomplishments have been impressive. Ninety-two percent of New Yorkers who enter a Robin Hood housing program don't return to shelters. Students who participate in Robin Hood–funded education programs have a 75 percent higher chance of passing the GED. And those entering foundation-backed job training programs are two times as effective in the workplace as those in other city-funded programs.[29]

True to its name, Robin Hood has been accomplishing its goals by taking from New York's rich to give to the poor, and it's managed to sprinkle some stardust on the performance-based work. The foundation is famous for its lavish galas, which attract performers ranging from Lady Gaga to Beyoncé and raise as much as $80 million in a single night.[30] It's also known for its who's who list of board members and donors that include bankers, hedge fund titans, and media and movie celebrities, such as Tom Brokaw, David Letterman, Sarah Jessica Parker, Uma Thurman, and Gwyneth Paltrow, to name just a few. It doesn't hurt that the foundation's board members and donors are some of America's uber-wealthy. Still, the foundation's deep Wall Street connections combined with its business focus have helped keep Robin Hood's legend alive. As board member and hedge fund manager Lee Ainslie told *Fortune Magazine*, "When we tell people their dollars are going to help their city, that every cent goes to those in need, that the board pays expenses, and that we hold our grantees accountable, you are talking about very powerful selling points."[31]

Another organization that's emblematic of the new-style philanthropy is the Omidyar Network (ON). Created by eBay founder Pierre Omidyar and his wife, Pam, ON is a philanthropic investment firm that invests in both nonprofits and for-profit organizations with a social focus. The bottom line for ON is social value, and the firm applies many of the best practices of the venture capital industry to scale up organizations that its leaders believe can benefit millions of people.

Much like a venture capitalist firm investing in a startup, ON invests in organizations based on factors such as innovation, scalability, and the viability of the business. Before making an investment, ON carefully evaluates the organization's management team, products and services, target market, competition, and financials.[32] When it invests in a for-profit company, ON considers the company's potential for positive social impact in addition to its financial metrics. For example, ON has invested in Amicus, a for-profit company that created a software platform to help nonprofits improve their fundraising efficiency and attract new donors. It is also investing in Bridge International Academies, a for-profit that's building schools across Africa to serve families living on less than $2 a day.[33]

When ON invests in a nonprofit, it encourages the organization to develop its own income stream so that it eventually becomes self-sustaining. For example, ON invested in DonorsChoose.org, an online charity that matches donors with public school teachers' requests for classroom materials. When donors support a specific project, the charity asks them to kick in an extra 15 percent to help pay for operating expenses and overhead. With these extra donations, the charity is now able to pay for its own operating costs without continued support from funders.[34]

Interestingly, DonorsChoose.org is itself pioneering a new type of philanthropy. Rather than relying on donations from traditional grant makers to pay for classroom supplies, the organization has turned to "crowdfunding," or small donations from hundreds of thousands of individual donors. One of the great things about DonorsChoose.org is that all of these donations are completely transparent. Donors can choose which classroom projects they fund. And once they make the donation, they receive photos of the project, a letter from the teacher, and information about how their dollars were spent.[35] This affords a glimpse at the model of the future. Rather than donating to some abstract cause, donors will know exactly where their investment is going and what impact it has. If we can better quantify results and tie them to the investment made, we can turn supporters of dreams into buyers of outcomes—an important shift.

Attracting More Capital into the Social Sector

Interestingly, the donor-as-investor mentality of these newer philanthropists is starting to carry over into the world of traditional foundations, some of which are reforming the way they work. Perhaps most notably, Judith Rodin, who since 2005 has been at the helm of The Rockefeller

Foundation, has been transforming the 100-year-old organization from a traditional grant maker into a social investor. Rodin led a thorough review aimed at figuring out how the foundation could double its impact. She cut back some of the foundation's long-term commitments, discontinued grants that weren't getting results, and began closely monitoring and tweaking the ones that were. "My rallying cry is to get leverage out of everything we do," she said.[36]

Unlocking Trillions of Dollars in the Private Sector

Under Rodin's leadership, The Rockefeller Foundation helped to coin the term *impact investing*, which strives to direct more money to social and environmental causes by unlocking vast amounts of money floating around the private sector. The idea is to bring together entrepreneurs, philanthropists, and private investors to develop new solutions that achieve the double bottom line of social impact and a financial return.

The Rockefeller Foundation first became interested in impact funding after concluding that private capital had to be unleashed to solve our environmental and social problems on the scale needed for success. Whereas philanthropy and government have billions to spend, the amount of money held by private markets is about $210 trillion, according to The Rockefeller Foundation figures.[37] "When you bring in private capital, you bring market forces to [bear on] philanthropic do-good impulses, and we feel that's really important," Rodin said. "It is wonderful to feel good, it is very important to do good, but it is most important to have impact."[38]

The most popular example of impact investing is the micro-financing industry, in which banks and other private institutions make small loans to low-income individuals to help them start businesses that lift them out of poverty. Working in partnership with others, The Rockefeller Foundation has been expanding this form of social investment to other social challenges, absorbing the initial risk and then attracting lower-risk investors once the projects get off the ground. For example, The Rockefeller Foundation helped create the New York City Loan Acquisition Fund, in which a group of foundations supplied the initial $36.2 million in high-risk capital for affordable housing projects. After the project took off and the risk was lessened, commercial lenders such as J.P. Morgan and HSBC stepped in to provide the second round of $190 million in funding.[39] Thanks to this program, 6,400 housing units have been created or preserved to date.[40]

The Rockefeller Foundation is also working to lift people out of poverty by bringing electricity to more remote parts of the world. As part of these efforts, it provided seed capital to d.light design, a for-profit solar technology company that manufactures solar lanterns and other affordable, energy-efficient lighting solutions and distributes them to people in developing countries. Although the $10 to $40 that d.light charges for its lighting products is expensive by developing world standards, it's a one-time expense that replaces the ongoing expense of buying kerosene, which in some regions costs $10 a month.[41] With the help of The Rockefeller Foundation and other investors, d.light has already provided lighting to 16 million people in poor regions of the world and is expected to bring reliable electricity to 100 million people by 2020.[42]

Using Private Capital to Solve Public Sector Problems

Impact investing is also starting to find its way into the public sector in the form of social impact bonds. As public sector funding becomes increasingly stretched, governments are searching for ways to leverage private sector capital to address social problems, with the goal of reducing government spending in the long term. In 2012, New York City became the first government in the United States to test social impact bonds, using them to fund a 4-year program aimed at reducing recidivism rates of adolescent men incarcerated at its main jail complex on Rikers Island. Before the program began, about half of all young men imprisoned at Rikers got new convictions within a year. In an effort to drop those numbers, Goldman Sachs will provide a $9.6-million loan for the program, and social services provider MDRC is designing and overseeing the program. Two nonprofit organizations, Osborne Association and Friends of Island Academy, are running the program, which offers counseling and education to an estimated 3,400 incarcerated adolescent men each year.

The idea behind social impact bonds is that private investors make a profit only if the initiative reaches a measurable target that's been agreed upon by the players in advance. In the case of New York City's jail program, Goldman Sachs will be repaid the entire $9.6 million if the program succeeds in reducing recidivism by 10 percent. If the rate drops more, Goldman could see its profits rise to as much as $2.1 million. But if recidivism fails to drop by 10 percent, Goldman could lose up to $2.4 million. New York City's Department of Corrections will pay Goldman if the program succeeds, taking the money it would have used to pay for a larger prison population at Rikers. Former New York City mayor Michael Bloomberg is also backing the program with money from his personal foundation, Bloomberg Philanthropies.[43]

Although social impact bonds are still in their infancy, interest in them is growing. To date, at least twenty-five such programs are being implemented around the globe, and more are being considered by local and state governments.[44] What's more, a bipartisan bill has been introduced in the U.S. Congress that would allow the federal government to allocate $300 million to support state-level social impact bond projects across the country. As social impact bond proponent Jay Gonzalez, Massachusetts's secretary of administration and finance, told the *New York Times*, "We've got to change the idea of, 'We just pay for stuff and hopefully get results.' The beauty of this is if they perform to get results, then we pay. If they don't, we don't pay."[45]

Although most governments are using these financial tools to solve problems such as reducing homelessness, recidivism rates, and the number of children in foster care, social impact bonds could just as easily be put to use to solve some of our most intractable environmental problems. Just think if the U.S. Department of Agriculture were to unlock billions of dollars in private capital to improve the quality of streams and rivers across the United States. And just imagine if that same outcome were tied to the profits of investment banks such as Goldman Sachs. You bet these projects would be successful, because failure wouldn't be an option—or at least not a repeatable one.

Capturing a Financial Return on Social Good

Aside from philanthropists and governments, momentum for impact investing is also coming from the private sector in the form of new investment opportunities that combine a financial return with environmental and social progress. Mainstream lending institutions from J.P. Morgan to Bank of America Merrill Lynch are devoting a small portion of their investment portfolio to impact funds. What's more, new investment houses with an exclusive focus on impact investing are cropping up, mainly in the United States and Europe. In the United States, the market has grown fivefold since 1995, with $3.74 trillion, or 11 percent, of total U.S. assets that are under management placed in sustainable investments.[46] At the same time, mainstream investments are becoming more transparent as organizations such as HIP Investor assign sustainability ratings to company 401(k) plans.[47] As Andy Sieg, head of global wealth and retirement solutions at Bank of America Merrill Lynch, put it, "We think impact investing is an idea whose time has come in mainstream wealth management."[48]

One investor who believes the market is ripe for impact investing is David Chen. Originally a partner at a traditional venture capital firm supporting the high-tech industry, Chen traveled to India to learn more about the micro-finance industry and became interested in how private investments could be used to address our greatest environmental and social problems. The result was Equilibrium Capital, a platform of funds that derives its value from social impact. The firm invests in hard assets ranging from green buildings to farmland to water assets.

Chen and his colleagues believe that sustainable practices are sound investments and will become even more so in the coming decades. For example, the firm created the ACM Permanent Crop, LLC, which invests in permanent cropland on the West Coast and in other parts of the United States. The fund, which to date has collected $250 million, supports farmers who are implementing sustainable farming practices to grow permanent crops such as citrus, blueberries, and grapes.[49] Because fruit trees and grapevines need only be planted once, crops such as these require an upfront investment with long-term yields. They are also high-value crops that consistently generate double-digit returns.[50] Equilibrium is placing its bets on the world's rapidly increasing population, which it believes will generate a larger middle class interested in eating fresh fruit and nuts, especially those that are organically grown. It's also betting that sustainable practices that reduce water and agrochemical use will be a competitive advantage in a world marked by permanent climate change and the growing scarcity of natural resources.

"I jokingly call our strategies Rudolph the Red-Nosed Reindeer," said Chen, who serves on The Freshwater Trust's board of directors. "Rudolph was made fun of by the other reindeer until that one stormy night. In the same way, impact investing is now a competitive advantage. In fact, it might be the only way to invest. We hope to influence the way people think, and we hope they will copy it." As John Fullerton so ably argues in "Regenerative Capitalism," it is precisely these examples of using finance to achieve the right relationship between economy and environment that will catalyze a fundamental shift in how we understand and use capitalism.[51]

Strengthening Our Impact at The Freshwater Trust

Like a lot of other nonprofits, The Freshwater Trust for many years had accepted the traditional funding model as a necessary headache of doing business in the nonprofit world. We wrote grants to a small army of funders and waited for their responses. And we submitted permit applications and waited for the green light. We knew these processes were inefficient. We knew they delayed good work, and we grumbled and fretted over it. But these were the hoops one had to jump through to make restoration progress—so we just kept jumping.

Our attitude toward traditional grant making changed abruptly about a decade ago after we began working on a small project to restore 1 mile of stream. The actual on-the-ground work would take about a week to complete, yet securing the necessary funding and permits took more than 3 years! It occurred to me: 3 years = 1 mile. For the first time, it brought the big picture into focus, and it wasn't pretty. In Oregon alone, 80,000 stream miles were in some need of help. Yet, in a good year, we were able to restore fewer than 600 miles. I did the math: 80,000 ÷ 600 = NEVER. The stark reality of the situation jarred me. True, we were staying busy, and we were scoring some wins. But at our current pace, our efforts were amounting to little more than a pothole in a massive, ever-expanding highway. Despite a lot of hard work, we simply weren't having a big enough impact.

Technology = Pace

As we pondered this realization, we decided to take a step back and assess what was impeding our progress, and one of our first conclusions was that we needed to speed up the grant and permit application processes so we could spend less time applying for grants and more time completing the on-the-ground work that was our raison d'être. Upon thoroughly researching the log jams, we realized that many of the different grant and permit applications often asked for the same information. What's more, we realized that there are just 16 ways to fix a river—not 6,000, not 60, but just 16. It wasn't a *science* problem that was holding us back but a *systems* problem. What was stopping us from making real progress was the fact that each government agency operates as a silo with its own set of procedures and requirements. In short, there was no standard approach to restoring rivers.

With this information in hand, we decided to automate these time-consuming and repetitive steps by putting them into in a single software platform. The result was our StreamBank software, a sort of TurboTax for restoration that combines all the regulations, design specifications, and funding requirements into an automated workflow. StreamBank has cut project completion times by 70 percent. Moreover, it has actually improved project quality by standardizing all the information and methods needed to get the work done.

To further speed up the restoration process, we worked with the Oregon legislature and the Oregon State Department of State Lands to exempt simple restoration projects from permitting requirements. We also persuaded the U.S. Army Corps of Engineers to agree to a "notice-based" permitting process for river restoration projects involving StreamBank, meaning that, because they had confidence in our ability to perform to their specifications, we effectively had a standing green light to move forward on certain types of restoration projects.

Finance = Scale

Technology gave us the pace we needed. But if we wanted to have a greater impact, we also needed to boost the scale of our restoration work by fixing many more rivers within a much shorter time frame. And that meant more financing—a whole lot more financing than what we were currently bringing in through the piecemeal system of applying for small grants. We realized that turning all those lines on the map of U.S. rivers as shown in figure 1.1 from white to blue would take trillions of dollars. And if we were serious, we needed to engage the multi-trillion-dollar private sector.

We've since found a source of funding for our conservation work that spans both the public and the private sectors while driving our mission forward. In many places along streams and rivers, the warm temperature of water prevents coldwater fish such as salmon from spawning. It also makes fish more susceptible to predators, parasites, and disease. To keep rivers at healthy temperatures, regulated entities such as municipalities, power plants, and industrial water users are all required under the Clean Water Act to limit the amount of clean but warm water they discharge into streams and rivers. Historically, these regulated entities have met these requirements by spending millions of dollars on structural fixes—namely, by building concrete water cooling systems. We decided to compete with these infrastructure fixes by offering a different solution: cooling water the natural way by planting trees that keep excess sun out of the watershed. Not only is it cheaper to plant trees, but trees offer a wide range of other ecosystem services in addition to cooling stream water, from stabilizing stream banks to filtering out pollutants before they reach the stream to providing food and habitat for wildlife.

To get the scale we needed, we worked with the nonprofit Willamette Partnership and the Oregon Department of Environmental Quality to turn these ecosystem services into quantified and approved credits that regulated entities could buy instead of building chilling towers—essentially creating a water temperature trade. So rather than building expensive gray cooling towers that require energy to operate, regulated entities can instead fund tree planting projects on farmland at the most critical points along the river.

A few years ago, we completed our first test project with the City of Medford in southern Oregon. City officials needed to find a way to cool the clean but hot water discharged from their wastewater treatment facility into the Rogue River so that it didn't interfere with the recovery of salmon. The city initially considered installing a concrete holding tank to the tune of $16 million to $20 million, but in the end they decided to work with us. In essence, the city bought water temperature credits derived from paying farmers within the basin to shift out of agricultural production into growing bushels of nature, if you will, through tree planting that shades and cools the watershed. By doing so, the city was able to offset its warm water discharge without installing an expensive cooling tower—meeting its Clean Water Act obligations for less than half the cost. The credits were developed using vigorous environmental accounting standards that translated restoration actions into quantified watershed cooling benefits measured in kilocalories. Willamette Partnership performed the mathematical conversion, certifying that the tree planting projects met the city's Clean Water Act obligations.[52] In the meantime, The Freshwater Trust worked with landowners and contractors to design and complete the restoration projects.

The City of Medford project demonstrated that water temperature trading programs can result in a win–win for the economy and the environment. By participating in our water temperature credit trading program, the City of Medford saved city ratepayers at least $8 million. It also added an estimated 100 temporary jobs to the local economy while providing a steady 20-year income for farmers and ranchers participating in the program.

The environment has benefited too. Over the coming years we will restore miles and miles of high-potential riparian area within the Rogue River Basin, securing a cooler watershed overall with direct benefits to the river. And, although the City of Medford is responsible for some 300 million kilocalories a day, we're completing more restoration projects than are needed, building in some 600 million kilocalories to ensure that the city is never out of compliance. Each credit project will be monitored for 20 years to ensure we get the intended results as the trees mature. And we've priced the credits beyond the actual cost of each restoration effort, giving us additional funding for future restoration projects. Because the margins are funneled back into more conservation, we can extend the mission impact even further.

With 200,000 regulated entities across the United States that have impacts that trading can address, the potential for watershed improvements are in the billions of dollars, which begins to right-size the effort to better match the scope of the problem at hand. Now that our water temperature trading program has been proven to work, we are working to replicate it—first throughout the Pacific Northwest and eventually across the entire United States. The critical next step is to standardize the biological, legal, and transaction mechanics so every project is like all the others.[53] We are also laying the groundwork for future environmental trading programs that solve watershed problems beyond water temperature, mainly preventing nonpoint source pollution such as nitrogen and phosphorus from flowing into rivers and, in the longer term, restoring imperiled salmon.

In the same way that a startup needs venture capital, we needed a large amount of initial funding to be able to replicate our water temperature trading program at scale. We eventually found that funding in the form of a long-term, low-interest loan from the David and Lucile Packard Foundation, a California-based family foundation that funds innovations in conservation and science. The foundation partnered with the Gordon and

Betty Moore and Kresge foundations to award us a $5-million program-related investment, which we are paying back at 1 percent interest over 7 years. It is an impact investment for them, in which the foundations are channeling their dollars into a social benefit as growth capital for us. The loan will cover our infrastructure and operating costs as we test and refine our water quality markets to ensure they're built with the rigor necessary to improve our watersheds in a lasting way. Together we're doing something that otherwise would not have happened—and that is what this game is all about.

In every case, the structure of the deal will more than offset any impacts to the environment. In order to make up needed ground, we expect the environmental gain of the restoration to exceed the impact from the industrial discharge on each and every deal, creating an overall win for the environment. In addition, efficiency and competition will keep costs low for municipalities, utilities, and industrial water users that participate in these deals. It will be incumbent on those doing the restoration work to identify sites that have the highest restoration potential at the lowest cost and look for ways to increase efficiency over time.

Water quality trading programs such as the one we entered into with the City of Medford are a pathway to the next frontier of conservation. As with DonorsChoose.org, all of our investors—whether they're private investors, foundations, or individual donors—will know exactly what their dollars are purchasing and what impact they're having. Rather than simply paying for the planting of trees or the establishment of a buffer bordering a river, investors will pay for the actual benefit—such as the specific amount by which solar loading into the watershed has decreased to benefit water temperature, the number of salmon restored, or the pounds per year of phosphorus reduced. In fact, they'll know exactly what they're getting because they'll pay for restoration projects only after they're completed and the benefits have been certified by a neutral third party. That way investors aren't doling out money toward an intangible problem that may or may not be solved. Instead, they are purchasing specific outcomes, with a measurable benefit for every dollar spent that will be monitored over a 20-year period to ensure lasting impact. By shifting from promises to quantified outcomes, we expect to drive significant improvement in the health of our rivers. We foresee a world in which all donors and investors become purchasers of environmental outcomes, if for different motivations. Where investors will build or purchase credits to sell to regulated entities, donors will purchase credits to retire for conservation purposes.

Accelerating the pace and scale of any work that benefits society represents a huge shift for social and environmental activists and the funders that support them. Yet the problems we face are too important for those opportunities not to be explored. Armed with ambitious goals, the right measurement tools, and twenty-first-century funding mechanisms, we can tackle our most pressing problems on a scale that matters. In the next chapter, we'll explore how we can put these ideas to work to tackle one of the most daunting river challenges: restoring the Colorado River, which is so vital to the U.S. economy and to our livelihood. As John F. Kennedy so aptly once said, "Change is the law of life. And those who only look to the past or present are certain to miss the future." The opportunity to make a difference is right in front of us; we just need to seize it and move full speed ahead.

CHAPTER 7

Lessons from an Aussie Water Shock

IF YOU'VE EVER BEEN to Las Vegas, you've probably seen the elaborate hotel fountains that shoot 500 feet into the air. You've probably swum in a Roman-themed swimming pool or played golf on a lush course kept green by hundreds of sprinklers. Maybe you've even ridden around in a gondola on an artificial canal modeled after the city of Venice.

Yet like a lot of things in Las Vegas, the seemingly endless supply of water is an illusion. The city, whose population has skyrocketed from 400,000 to two million residents over the last decade, relies on Lake Mead for 90 percent of its water.[1] But the problem is that Lake Mead is running dry.

Created by the construction of the Hoover Dam on the Colorado River, Lake Mead was filled to the brim less than 2 decades ago. But in 2000, the nation's largest reservoir started to drop, and today it is about 40 percent full.[2] If the water level drops below 1,000 feet above sea level, Lake Mead will be too low to carry water through the two tunnels that supply Las Vegas, Los Angeles, and several states. Yet that's exactly what's happening. Today, Lake Mead stands at about 1,100 feet above sea level. If we continue to gobble up water at the current rate, the water level will sink far below 1,000 feet, with a 50 percent chance of becoming a "dead pool" by 2036.[3]

Anticipating the crisis, the Southern Nevada Water Authority has been busy drilling a new $1-billion tunnel to suck up the last remaining water as the water level continues to drop. In addition, Las Vegas wants to build a $15.5-billion pipeline that would pump groundwater from an aquifer 260 miles away in rural Nevada. But environmentalists are challenging the proposal, and so far the courts have agreed with their argument that the project would harm habitat for trout, birds, deer, and elk.

In the meantime, Las Vegas has taken significant steps to cut back its water use. It's subsidizing the cost of water-efficient appliances. It's offering a rebate program to convince residents to rip out their grass lawns.[4] And it's been recycling almost all water used indoors by its 40 million annual visitors and returning it to Lake Mead.[5]

As admirable as these actions are, they've been no match for the scope of the problem, and Las Vegas continues to lose its gamble with water. "The situation is as bad as you can imagine," said Tim Barnett, a climate scientist at the Scripps Institution of Oceanography. "It's just going to be screwed. And relatively quickly. Unless it can find a way to get more water from somewhere Las Vegas is out of business."[6]

Las Vegas isn't the only place that's hurting from a diminished Colorado River (figure 7.1). For 6 million years, the 1,450-mile river flowed from the snowy Rocky Mountains in Colorado southwest through seven states, emptying 14 million acre-feet of freshwater into the Sea of Cortez in Mexico.[7] Yet today, the Colorado rarely reaches the ocean, creating an ominous situation for the 40 million people in the American Southwest and 3 million people in Mexico who depend on the river system for their livelihood. The situation is also foreboding for agriculture, which sucks up three quarters of the river's water to irrigate 3.5 million acres of cropland.[8]

Unfortunately, circumstances are expected to become even more dire

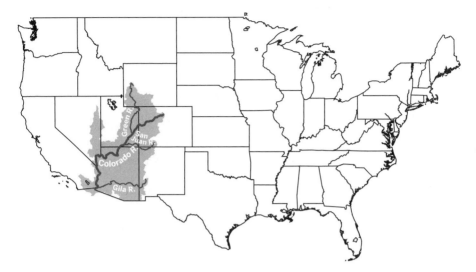

Figure 7.1. Today, the Colorado River rarely reaches the ocean, creating an ominous situation for the 43 million people who depend on the river system for their livelihood. *(Credit: The Freshwater Trust.)*

as population in the region grows and climate change causes longer and deeper droughts. The population of the American Southwest is expected to almost double by 2060, with another 1 million people moving to Las Vegas alone.[9] Within the same time period, climate change could reduce the river's flow by as much as 35 percent.[10]

Despite the warning signs, Las Vegas continues to roll the dice by allowing the building boom to continue. "How foolish can you be?" said Barnett. "It's the same fatal error being repeated all over the Southwest—there is no new water."[11] Right now, our water math isn't working.

A Similar Scenario 9,000 Miles to Our South

All signs are that the American Southwest is heading for a water crisis of huge proportions. Yet, as threatening as the situation is, there is a way to address this problem, and interestingly, a precedent exists. Nine thousand miles away, Australians have been experiencing their own water crisis. We could go a long way toward restoring the Colorado if we learned from their experience and applied the same lessons here at home.

So what's the situation in Australia, and how does it compare to what's happening here in the United States? Like the Colorado River, the 1,600-mile Murray–Darling River is critical to Australia's livelihood. It's the major domestic water source for more than 1 million people in southeast Australia, and it supplies 65 percent of all water used for the nation's agriculture.

As in the Colorado River Basin, climbing temperatures and plummeting rainfall have been reducing water flows. The average temperature in the Murray–Darling Basin has increased 1.6 degrees Fahrenheit since 1950, with temperatures reaching as high as 120 degrees Fahrenheit during the summer.[12] At the same time, rainfall over the past 15 years has decreased precipitously.

And, as in the Colorado River, the demand for water has been outstripping supply as population continues to grow. Today, population in the region has climbed to 2.1 million, and water use has grown more than fivefold since the 1920s.[13]

Yet, unlike the United States, Australia has reshaped its attitude toward water, discarding the old view that water is limitless and replacing it with a comprehensive set of solutions for managing its use. Unfortunately, it took a crisis to get there. But today Australia is tackling its water problems head on, and it's taking a quantified approach every step of the way.

Australia's Water Shock

As is the case with the Colorado, Australia's water problems didn't just happen overnight. They are the result of 150 years of loose management that have allowed the overuse of a resource that's turned out to be less plentiful than once assumed. Even in Australia's dry ecosystem, water until recently had been taken for granted.

One of the driest regions of Australia, the Murray–Darling Basin has always had unpredictable rainfall patterns that include prolonged droughts followed by substantial flooding. Such patterns disrupted the rhythms of early economies, and European settlers compensated by channeling water from rivers to irrigate their land and store extra water for use during the drier times. To do that, they dammed, straightened, and controlled the waters of twenty-three river valleys to ensure they had enough water for human uses. The situation isn't that different from that of the Colorado, which has fifteen dams on its main stem and hundreds more on its tributaries.

The increased water use eventually turned Australia into a booming agricultural economy, yet in recent years climate change has brought about longer and more devastating droughts. The worst drought in Australia's recorded history, the "Big Dry," starved the region for water, cutting river flows to less than half of what they once were.[14] With little to no water, ranchers were forced to slaughter or sell off their cattle, and farmers of water-intensive crops such as rice and cotton let their fields lie fallow. In the end, more than 10,000 farmers have been forced off the land,[15] and suicide in the region has been at twice the national average.[16]

The combination of increased water use and prolonged drought has taken a toll on the ecosystem, too. The Australian government has designated twenty of the twenty-three major river valleys as in "poor" or "very poor" ecological condition.[17] And like the Colorado, the once mighty Murray River doesn't always reach the ocean. In fact, it does so now just 40 percent of the time. There are other problems, too: Wetlands and floodplains have been lost to European settlement, river salinization has been growing in frequency, and native fish populations are now only 10 percent of what they once were.[18]

Yet it took an economic catastrophe for Australia's water crisis to make international headlines: In 2008, the country's million-ton rice crop failed,[19] doubling rice prices and sparking food riots in 34 countries.[20] The situation demonstrated firsthand the ripple effect a water crisis in one food-producing region could have on the rest of the planet. No industrialized nation since the American Dust Bowl of the 1930s has experienced more damage from drought and water scarcity in its prime food-growing region,[21] and the effects were being felt around the globe.

Within Australia, nearly everyone was suffering, including residents of urban areas. In South Australia's capital city of Adelaide, for example, residents were living in a constant state of water shortage. With a reliance on the Murray River for up to 90 percent of its water during periods of low rainfall,[22] thousands of the city's 1.3 million residents found their water supplies suddenly cut off, forcing them to fulfill their basic needs from water tanks trucked in by the municipality. "Things are so bad that we wash our clothes in the cattle's water troughs and sometimes have to scoop water from the pond at the golf course," said Adelaide resident Martha Christian. "What other choice do we have?"[23]

Although the rains that began in 2010 provided a temporary reprieve, by 2014 the drought had returned. The persistent hot, dry weather has been turning rural communities into ghost towns and pushing whole growing sectors including rice, cotton, and citrus producers to the brink of collapse.[24] Many have predicted that the pattern of less rainfall and a hotter, drier climate is "the new normal" and have been urging Australia to adapt. "If the sort of climatic regime we've had in the past couple of years becomes a feature of the future, it's pretty clear we don't have the volume of water available that we've had in the past," Wendy Craik, former chief executive of the Murray–Darling Basin Commission, told *BBC News*. "Clearly the basin is not going to be the same."[25]

A Coordinated Response

When the demand for water outstrips the supply, the situation quickly escalates into a crisis. And when that happens, even reactive governments are forced to respond. In the case of Australia, the national government eventually did respond, and did so in spades. Realizing that both its economy and its largest river system are in jeopardy, Australian policymakers have begun to implement bold reforms that transform its relationship to water. Like the Colorado and other major rivers throughout the United States, the Murray–Darling Basin had been managed by a kaleidoscope of jurisdictions, making effective management a challenge.

Until recently, water in the basin had been regulated largely by the century-old River Murray Water Agreement of 1914, which gave multiple states within the basin the authority to store 3 years' worth of water in reservoirs and otherwise allocate water in the region as they saw fit. About 90 percent of the water had been allocated to farmers, with the rest set aside for the towns growing up around these rural areas. The idea was to encourage the development of agriculture by providing as much water as possible. Yet, as it became clear Australia was entering a new era of permanent water scarcity, policymakers began to reevaluate this approach.

The country took an important step in 1995, when for the first time it capped the amount of water that could be siphoned off from the basin.[26] In 2004, the Council of Australian Governments, an intergovernmental forum that institutes national policy reforms, extended these reforms by implementing the National Water Initiative, which called for a more cohesive national approach to the way Australia manages, measures, plans for, prices, and trades water.[27] In doing so, it established a National Water Commission charged with assessing the progress of state and federal governments toward these goals.

Three years later, the federal parliament enacted the Water Act 2007, aimed at remedying the overallocation of water in the Murray–Darling Basin by bringing it under federal management. Considered the most extensive Commonwealth intervention in water resource management in Australia since federation, the Water Act 2007 allocated $9 billion (USD) to a wide range of initiatives, including the development of a basin plan that caps the total amount of surface and groundwater that can be withdrawn from the river system.[28]

The legislation also established the Murray–Darling Basin Authority (MDBA), a single government agency responsible for managing water across the entire basin. To give the MDBA the authority it needed, the four states and one territory touched by the basin were required to delegate some of their water management powers to this new agency. As former Australian prime minister John Howard declared, "The old way of managing the Murray–Darling Basin has reached its used-by date."[29]

The new approach to managing water drew angry confrontations from farmers, including the public burning of a draft plan. It also led to the eventual resignation of the MDBA's original chairman.[30] Yet the federal parliament persisted and in November 2012 passed a basin plan into law. The old adage of "no pain, no gain" applies here. This was a *radical* reset of more than a century of vague management, representing a vastly improved water game plan for the coming century. Often changes of large magnitude are accompanied by political pain and unrest. But in the long term, that pain will be lessened by a management plan that reliably and fairly allocates limited water supplies.

The ambitious plan doles out water based on how much water is actually in the river basin, measuring water levels on an ongoing basis and prorating water as circumstances change. It imposes new, measurable water quality standards. And it includes a monitoring and evaluation program that summarizes the region's progress achieving its goals. The plan, which is being phased in over a period of 7 years, is aimed at balancing the water needs of communities and industry while also protecting the ecological health of the region's watersheds. Craig Knowles, now chair of the MDBA, equates Australia's water reforms to the start of a long journey. "We need to recognize that doing nothing is no longer an option," he told me in an interview. "So we have to make a start—and through the process of ongoing management, monitoring, and evaluation—have the willingness to adapt as new information comes online."

A Quantified Approach to Water Management

Of course, across human history promises of good action have had a checkered past. But Australia's water realities will accept little else. The hard decisions and steps taken to date speak to the commitment of a country determined to transform its water future. Arguably, no other country has laid out such an ambitious approach to managing a watershed. In particular, what makes the Murray–Darling a model for the rest of the world is that it is taking a quantified approach to water reform. From the beginning, the goal has been to use the "best and latest scientific, social, cultural and economic knowledge, evidence and analysis."[31] In fact, to some degree the plan touches on all five principles of quantified conservation. "The important thing to understand is that our plan, by law under the Water Act, must be validated in a scientific method," said Knowles. "We draw from 120 years of climate data, and all of our work is peer reviewed."

Sizing Up the Situation

To develop an effective plan, the MDBA's first step was understanding the baseline from which it was starting. In the same way that a financial budget helps determine how much money a person brings in and how much he's spending, the MDBA needed a water budget. And to create a water budget, officials needed to know both how much total water there was to work with and how much water was being used. Using data and analysis, the MDBA was able to determine that the average surface water inflow was 8.7 trillion gallons per year, and the average groundwater recharge rate was 7 trillion gallons per year. They also determined that the total amount of surface water being withdrawn each year was 3.6 trillion gallons, and the total amount of groundwater withdrawals was 449 billion gallons per year.[32] Given seasonal changes in water levels and other variables, the straight water math simply cut things too close. Such little room for error had led to unreliable water supplies for agricultural production, and it also held consequences for environmental health.

The MDBA also needed to understand the current conditions of its watersheds—including factors such as oxygen levels, volumes of blue-green algae, and salinity levels. Once it had this information, the authority was able to better manage water quality on a more integrated basis. For example, salinity poses a major challenge in the Murray–Darling Basin. In hot, dry parts of the basin where water evaporates quickly, salts from the water are left behind, reducing crop yields, damaging everything from orchard trees to infrastructure, and seeping into drinking water. Altogether, the MDBA calculated that an estimated 2 million tons of salt had accumulated in the river system.[33] Knowing both the total amount of salt that the basin dumps into the ocean each year and the salinity levels at various reporting sites throughout the basin, the MDBA has been able to set measurable targets and evaluate whether it is achieving those targets over time.

To better understand the state of its watersheds, the MDBA also took into account both the current and the future impact of climate change, concluding that its plan needed to incorporate strategies to cope with more violent and longer-lasting storms as well as longer, more severe droughts than in the past. Using modeling, the authority determined that by 2030, water availability could decrease by as much as 27 percent under the most extreme dry scenario.[34] With the ability to predict how climate change might affect future water availability, the MDBA has the information it needs to plan for the future.

Changing the Game through Bold Outcomes

Once officials understood the baseline from which they were starting, they could then figure out what they wanted to achieve. To that end, the MDBA established four outcomes: improved ecological health, better water quality, more reliable water management policies, and communities that are better adapted to dealing with water scarcity.[35] Note that these outcomes don't just call for Australia to stop more bad things from happening. They change the game entirely by demanding large-scale improvements. These are the kinds of bold outcomes that lay the foundation for real progress.

For a plan to be successful, outcomes must be measurable. And at the heart of the basin plan—and its most controversial component—are "sustainable diversion limits" that dictate how much total surface and groundwater can be taken from the two major rivers and their tributaries without jeopardizing environmental health. These limits establish a water budget that is within the carrying capacity of the physical environment. The plan calls for water access rights to be cut by 726 billion gallons per year—a 20 percent reduction over 2009 levels—so that this water can be returned to the rivers. In addition, it establishes 878 billion gallons as the annual cap for total groundwater withdrawals. To leave enough water in-stream, the government is buying back water licenses from irrigators and investing in water-saving technology.

Embracing Innovation to Reduce Water Use

By setting bold outcomes, the MDBA has effectively demanded the development of innovations to help bring about this new reality. To that end, the government has committed a whopping $4.6 billion (USD)—roughly one third of what the U.S. government pays out in annual farmer crop insurance subsidies—to improve efficiency of rural water use in the Murray–Darling Basin, with the savings to be shared by irrigators and the environment.[36] Some of this money is being spent on grants to help farmers improve irrigation efficiency by laser-leveling paddocks, modernizing drip irrigation systems, and installing soil moisture monitoring equipment.[37] Another portion of these funds is being used to help states install better irrigation infrastructure, such as pipelines to replace open ditches, where water is lost through evaporation.[38] And some of this money is being spent on projects that encourage cities to modernize their water infrastructure. Combined, these projects are expected to save 159 billion gallons of water per year.[39]

Projects like these are examples of how we can work smarter to conserve water in an era of permanent water scarcity. The opportunities for innovation are endless. By setting clearly defined outcomes, the Australian government is jump-starting the market for water-saving technologies, and both the private and public sector are seizing the opportunity.

Adapting to the Reality at Hand

Another key to the success of the basin plan is a comprehensive monitoring and assessment program, along with transparent reporting, to evaluate whether the plan is working. In addition to compiling existing data from a wide range of organizations, the MDBA has supplemented these data with its own modeling tools. Using models, for example, it examined how the ecological health of the basin fared over a 100-year period. It is also updating its groundwater models to determine flows and recharge rates. It is using models to predict how agriculture and industry are likely to use water based on past experience. And it commissioned a study on how changes in water allocation levels would affect the agricultural industry and individual users in the region.[40] The MDBA has committed to making its data available for public scrutiny while using it to inform future decision making.

In terms of compliance, the MDBA is developing an annual volume of "permitted take" for each irrigator that will vary in response to climate changes, river flows, and other factors. At the end of each year, the authority will audit whether actual water withdrawals matched what was permitted. The MDBA plans to publish an annual report charting water use. And every 5 years, it will review both the environmental watering plan and the water quality and salinity management plan.

Not only is the MDBA putting data and analysis to use to inform its work, but it is taking an adaptive management approach in which changes will be made as more information is gathered and officials evaluate the success of the plan. "It's very easy for the players to think the problem has been dealt with and move on to the next subject when, in truth, this is a never-ending, constantly-changing natural landscape that needs to be monitored and evaluated," Knowles said. "We have to have adaptable plans to reflect the changes as and when they occur. And the overarching framework needs to be consistently reinforced and reevaluated." This is key to the success of any water management plan. It's not enough simply to put a plan into effect. We must also monitor our success and make the required changes when the plan isn't working.

Encouraging Water Trading across the Basin

As part of the basin plan, the MDBA is also improving its market for water trading. First introduced in the Murray–Darling Basin in the late

1980s, water trading allows people with too much or too little water to trade their water access rights. Once a water budget is set with minimum river flows to protect the environment, water allocated for human uses can be traded. A person who needs more water than they have can buy it from another person. Likewise, a person who doesn't want to use all the water they have can sell it—either permanently or for a limited amount of time.

Already, water trading in the Murray–Darling Basin has benefited Australia's economy. By selling water, farmers have been able to better manage debt and cope with drought, and many have shifted to less thirsty but higher-value crops. By buying water, some farmers were able to maintain their production or keep permanent plantings alive during the long drought. Indeed, nearly one third of dairy farms, 20 percent of broadacre farms, and 23 percent of horticulture farms have been buying and selling water on a temporary basis (figure 7.2).[41] And in 2008–2009 alone, water trading increased Australia's gross domestic product by $192 million (USD).[42] As with any market, values and participation will fluctuate over time, but directionally, the proper pricing of water within the carrying capacity of the environment is the right way to go.

At the same time, water trading benefits the environment. For example, the federal government has spent about $2.6 billion (USD)[43] to buy about 396 billion gallons[44] of water back from voluntary sellers to ensure that enough water remains in-stream to protect the health of rivers, floodplains, and wetlands. This water is literally dumped into places along the rivers that need it. For example, between July 2012 and June 2013, the government returned 264 billion gallons of water to improve river flows in the southern Murray–Darling Basin.[45] The water was strategically applied to locations that promoted fish movement, restored wetlands, and improved water quality by ensuring that salt and other nutrients were flushed out to the sea.[46]

Yet, although water trading has been increasing within the Murray–Darling Basin, it still isn't being used by the majority of farmers.[47] A major restriction has been the separate management of water trading within each state. Although the Murray–Darling Basin extends across four states and

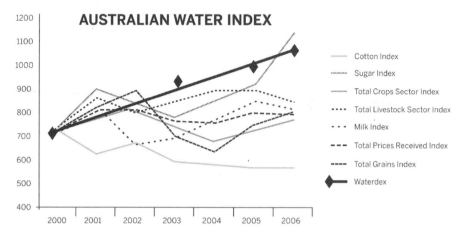

Figure 7.2. When Australia valued and traded its scarce water supplies, the price of water shot up, beating the indexed price of several commodities. As a result, many farmers in the Murray–Darling basin changed to less water-intensive and higher-value crops. *(Credit: The Freshwater Trust, adapted from Michael D. Young, "The Effects of Water Markets, Water Institutions and Prices on the Adoption of Irrigation Technology," figure 7, p. 17. For long-run water price index [Waterdex] and indices of the value of agricultural commodities developed by the Australian Bureau of Agricultural Economics, see Psi-Delta, http://www.psidelta.com/waterdex .html.)*

one territory, water trading had not been allowed across state boundaries, limiting the number of water trades that could take place. As Knowles explained it, "Farmer Brown on one side of the river might need water and Farmer Green on the other side might be happy to sell some water, but the rules, because of state jurisdictional boundaries, prevent a transfer that in any other circumstances would be a logical and natural trade exchange."

The new rules are designed to overcome that limitation by allowing water to be traded across state lines within the same river basin. They also provide a consistent framework that includes the same set of trading rules regardless of where the water is traded.

Australia's water market is helping the economy by providing farmers a more flexible way to cope with drought and water scarcity. At the same time, it is benefiting the environment by creating opportunities for both the government and conservation groups to buy back water and return it to the rivers to restore the ecological health of the Murray–Darling Basin. And, in the long term, Australia's water trading market could bring additional money into river restoration by encouraging investment in conservation finance.

Criticism of the Basin Plan

Predictably, the plan has drawn its share of criticism from both users whose allotments got curtailed and conservation groups that wanted even more for the environment. When the final sustainable diversion limits were announced, farmers claimed the plan would devastate Australia's rural economy. "For the communities, the family farms and the local businesses across the basin, the result is more than disappointing—their very futures are on the line," Matt Linnegar, chief executive of the National Farmers Association, told *The Guardian*. "The impact of this will be job losses, closure of family farms, hardships for regional communities and increases in fresh food prices."[48]

At the other end of the spectrum, environmental groups argued that the volume of water returned to the basin wouldn't go far enough to protect river health. "This draft plan fails the river, regional communities and our national interest, because it doesn't do enough to flush the salt out through the Murray mouth, revive dying wetlands and keep the country's lifeblood—the Murray–Darling—flowing," said Paul Sinclair of the Australian Conservation Foundation.[49]

Notably, the respected Wentworth Group of Concerned Scientists, an independent group of scientists, accused the MDBA of "manipulating science" in an "attempt to engineer a predetermined political outcome."[50] The Wentworth Group argued that the 726 billion gallons that will be returned as surface water in the basin is below the range of 792 billion to 2 trillion gallons the agency initially concluded was needed to restore a healthy environment.[51] It accused the MDBA of deciding on this limit first, and then tailoring the science to fit its predetermined conclusion.

The Wentworth Group also noted that, while the MDBA reduced the sustainable diversion limit by 726 billion gallons, it established a limit for groundwater that exceeds current withdrawals by 687 billion gallons. This, too, deviated from the MDBA's original conclusion that groundwater withdrawals needed to be reduced by 26 to 59 billion gallons per year. The Wentworth Group concluded that the current plan "falls well short of returning the volumes of water that science has shown are required for a healthy river."[52]

If the discrepancy resulted from a revision in modeling, that's fine. Not so fine if it was a concession to pressure. Either way, there are bound to be stumbles in the early phases of such a shift in paradigm. The beauty of a quantified approach is that we can measure whether water reforms to the Murray–Darling Basin are working. Using existing technology, we can determine whether fish populations are rebounding. We can measure whether salt and other nutrients in the water are declining. And we can establish whether the volume of water returned to the basin provides the river flows needed to restore wetlands, floodplains, and wildlife to the region. In the end, the risks associated with disregarding realities in a century of permanent scarcity will quickly show themselves and demand correction. That's the value of measuring progress toward a predefined outcome.

Learning from Our Neighbors Down Under

So what can we learn from Australia's approach? A national water act like the one Australia enacted would certainly make it easier to manage our rivers. Yet with a less than functioning U.S. Congress, it's doubtful that's going to happen. In the meantime, however, the seven states touched by the Colorado River could take a coordinated approach by relinquishing some of their individual authorities and wants to an entity similar in construct to the MDBA that would manage water across the entire Colorado

Basin. By delegating some of their water powers to a regional entity, states would allow water to be allocated more realistically, providing enough water for ecological health, driving higher-value crops, catalyzing smart trades, and still leaving enough water for Mexico.

States bordering the Colorado have made some attempts to allocate water in a coordinated way. Most notably, they signed the Colorado River Compact, which in 1922 divided the seven states touched by the Colorado into an upper and lower basin, each of which was allotted 7.5 million acre-feet of water per year—to be split up between the states in each basin. In addition, another treaty signed in 1944 allocates 1.5 million acre-feet from the river to Mexico.

A Water Budget Based on Actual Water

The Colorado Compact purports to divide up water in a rational way, but when push comes to shove, it's about each state getting its own. And historically, every time the region has started to put on the brakes, it's been saved by a wet year. That ain't gonna last. Until we take a reality-based approach to water, this region is heading for a crash that we will all watch in real time.

What's more, the Colorado Compact made water allocations based on average annual river flows collected from 1899 to 1920, which happened to be an unusually wet period. Not only was the compact based on water flow assumptions that were overly optimistic at the time, but rainfall in the region has dwindled significantly over the past century because of climate change. The result is that more water is being allocated than actually exists. In a nutshell, our water math is really water fiction.[53]

In an attempt to address the shortfall, the U.S. secretary of the interior in 2007 set guidelines that reduce the amount of water states would receive based on light, heavy, and extreme shortages, determined by measuring the amount of water in Lake Mead.[54] And in 2012, the United States and Mexico signed a new agreement that adjusts the amount of water Mexico is to receive in drought and rainy years.[55]

Although these attempts are a step in the right direction, they take into account only our immediate economic needs, short-shrifting the environment completely. Any serious effort to manage water volume starts with a water budget based on the *actual* water that's in the basin. It must include a minimal environmental flow that leaves enough water in the river for a functioning environment. And it needs to include limits for both surface water and groundwater because the two are interconnected. The math simply has to work. Hope is not a strategy.

Prorated Water Rights

Another lesson we can learn from Australia is the use of data and analysis to help us achieve the results we seek. By measuring snowpack, rainfall, and other factors, we can accurately model water flows for the coming year to create a realistic water budget that starts with minimum environmental flows to protect the river system. Above those minimums, we can then prorate the volume of water each water rights holder can withdraw, allowing both farmers and urban users the time to adjust.

We can also harness data and analysis to plan for the future. We can model the impact of different water allocations on both the economy and the health of the watershed. And we can use modeling to predict how climate change will affect water supplies over time.

Additionally, data and analysis can help us to determine whether our attempts to restore the Colorado are actually working. As with the Murray–Darling, the situation with the Colorado River is constantly changing. By evaluating and reporting on our results and adapting the plan as needed, we can make sure we're obtaining success on an ongoing basis.

Water Conservation

We should also follow Australia's lead in embracing innovation and technology to conserve water—and do so on a much broader scale. Once sustainable diversion limits are set that preserve the health of the river, innovation will follow. Just as Las Vegas has responded to water scarcity

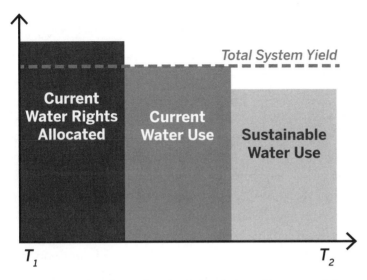

Figure 7.3. Water rights allocated in the Colorado River Basin exceed the total physical amount of water in the system, and current use far exceeds that which is sustainable. *(Credit: The Freshwater Trust., adapted from "Colorado River Basin Water Supply and Demand Study." U.S. Department of Interior Bureau of Reclamation. December 2012.)*

by recycling the water used by tourists, we can work to conserve water in a serious way in urban and rural areas throughout the basin.

Conservation can come in the form of incentives that encourage urban users to conserve water while motivating municipalities to modernize their water infrastructures. It can also come in the form of grants that help farmers improve irrigation efficiency or change to less thirsty crops in dry areas. Just imagine if we took a sizable chunk of the annual $14 billion that the U.S. government spends to insure farmers against crop loss and instead used it to improve water efficiency? You can bet there'd be a boom in water-saving technologies.

A Basin-Wide Trading Program

Finally, we should set up a basin-wide water trading program that encourages those who own water rights to conserve water and sell their surplus supplies to those who need it. As Australians have learned, the most efficient water market is a single program built on a single set of mechanics that encourages a large volume of trading to take place. By implementing a basin-wide water trading program, the seven states that border the Colorado River would offer rural and urban areas a more flexible way to cope with water scarcity. At the same time, they would provide a standardized way for governments and environmentalists to buy back water and return it to the river to preserve its long-term ecological health.

Quantified conservation makes it possible for all these improvements to happen, paving the way for a Colorado River system that's managed with far greater precision and sophistication. By taking a quantified approach, we can provide the right incentives to conserve limited water supplies while ensuring that water is properly valued.

A Source of Knowledge and Hope

The jury is still out as to whether Australia can successfully implement its impressive list of reforms. Yet one thing is clear: Australia's quantified approach to solving its water crisis serves as both a source of knowledge and a source of hope for the Colorado River and other rivers across the United States as freshwater becomes scarcer.

Australia's bold water reforms demonstrate that a new era of water management is possible. For that to happen, a huge new paradigm will be needed on many levels. Yet this shift is as doable as it is inevitable. The fact is, drought is coming at us in a brand new way, and its persistence will lay bare the shortcomings of how we manage our water. Having the resilience to deal with it will make the difference between failing and a thriving future. The choice is ours: We can wait for a Pearl Harbor moment, which will inevitably come and leave us no choice but to act with great haste, or we can begin the needed reset even as water in the United States advances to that crisis point.

As we'll see in the next chapter, it's not just water scarcity that we must address but also the declining quality of our freshwater. As we dump more agrochemicals in our streams and rivers, we've been creating a situation that has long-term ramifications for both the environment and the U.S. economy. Yet here, too, there's a better way.

CHAPTER 8

Getting Clear on
the Big Muddy

NANCY RABALAIS has spent her life sounding the alarm bells about the massive dead zone at the mouth of the Mississippi River. Every year, she and her team of researchers at Louisiana Universities Marine Consortium measure areas of low oxygen that are suffocating plant and marine life in the Gulf of Mexico, issuing a press release about the changes they observe. Since she began monitoring the dead zone in 1985, it has more than doubled to 6,700 square miles—an area larger than Connecticut.[1]

To get folks to understand the horror of living without oxygen, Rabalais likens the phenomenon to "stretching a sheet of plastic wrap from the mouth of the Mississippi River west to Galveston, Texas, and sucking out all the air."[2] That fiction would be no good for humans, and the reality is that it's no good for fish. Once teeming with fish and other marine life, the area has essentially been turned into a biological desert, an ocean wasteland where life is absent. "You can swim and swim and not see any fish," Rabalais said. "Anything that can't move out eventually dies."[3]

The dead zone is a seasonal phenomenon that forms each spring. Rabalais and her team have linked the dead zone to vast quantities of nitrogen and phosphorus flowing down the Mississippi River from sources such as fertilizers, animal manure, sewage runoff, and industrial waste. As these nutrients settle into the mouth of the Gulf, they form algae, which starve living things of oxygen. Over the past 3 decades, the amount of nitrogen flowing into the Gulf has increased by up to 300 percent, most of it from agriculture.[4]

When their early findings were published in *BioScience* in 1991 and *Nature* in 1994, Rabalais and her team drew national attention. Yet she quickly realized it wasn't enough to simply write about her results, so she overcame her stage fright to speak before audiences ranging from middle school students to the U.S. Congress. Her message: The Gulf dead zone is living proof that everything's connected and that everyone must take responsibility for their actions. As she wrote in one article, "We ALL live downstream from our fellow world co-inhabitants and upstream from our inputs."[5]

An American Icon

The world's third largest river, the Mississippi is one of America's most cherished icons. It's a symbol of freedom and escape in some of our greatest novels. It's the serene backdrop in some of our greatest art. And it's a source of inspiration to some of America's finest jazz and rock-n-roll musicians. The Mississippi's influence on American culture is almost as long as the river itself, which run 2,340 miles from its source at Lake Itasca in Minnesota through the center of the continental United States to the Gulf of Mexico (figure 8.1).

Aside from its role in American culture, the Mississippi River also has massive importance to the U.S. economy. Draining thirty-one states and 40 percent of the total landmass of the United States, the Mississippi River Basin is the agricultural heartland of the United States. More than half of all goods and services consumed by Americans are produced with water

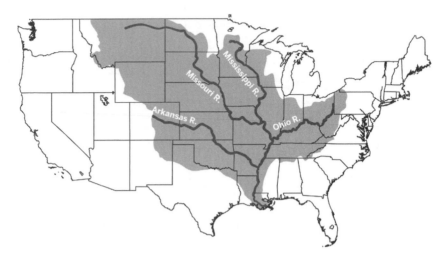

Figure 8.1. The Mississippi River Basin drains thirty-one states and 40 percent of the total landmass of the United States. *(Credit: The Freshwater Trust.)*

that flows through the Mississippi River and its tributaries. The region generates more than $54 billion in corn, grain, livestock, poultry, cotton, sorghum, soy, and other agriculture products and 92 percent of our farm exports each year.[6] The river is also an important artery of commerce, with barge traffic moving thousands of tons of agricultural fertilizers, coal, petroleum products, construction materials, steel, and industrial chemicals to river ports throughout the Midwest. And it serves as a prime tourism attraction, generating $1.2 billion in annual revenue from sport fishing, waterfowl hunting, and other recreational uses in the upper basin alone.[7]

In addition to its economic prowess, the Mississippi plays a star role in the health of our environment. Roughly 60 percent of all North American birds use the Mississippi River Basin as their migration corridor. The Mississippi is also home to 25 percent of all fish species in North America, at least thirty-eight different types of mussels, and fifty different species of mammals. What's more, 18 million people in more than fifty cities rely on the Mississippi for their daily drinking water.[8] In short, a functioning Mississippi River is of daily importance.

One of America's Most Endangered Rivers

Yet much has changed since Mark Twain declared that the Mississippi "will always have its own way." Today, humankind is having its way with the Mississippi, turning it into a poster child for how we abuse our natural resources. As increasing levels of nutrients flow into the Mississippi at numerous points along the river, the river's water quality has been steadily eroding. Nitrogen levels in the Mississippi have skyrocketed from less than 1 million tons at the advent of the Green Revolution to about 13 million tons in the mid-2000s, creating far more pollution than the river can absorb.[9] The consequences have been devastating. Today, 39 percent of streams in the Mississippi River Basin have high levels of nitrogen, and 32 percent have high levels of phosphorus, choking the diverse river life that calls the Mississippi home.[10]

By far, the biggest sources of these pollutants are agricultural fertilizers.[11] Eighty percent of all corn and soybeans produced in the United States are grown in the Mississippi River Basin. And corn is an especially fertilizer-intensive crop, accounting for more than half of all fertilizer applied to crops in the United States. Runoff from corn and soybean fields into the Mississippi River Basin makes up more than half of the nitrogen pollution and one quarter of the phosphorus pollution that ends up flowing into the Gulf of Mexico.[12]

The region is also where most of America's factory-scale meat farms are concentrated, including beef cow, dairy, hog, chicken, and egg farms. Together, animal waste from these feedlots accounts for 5 percent of the nitrogen and 37 percent of the phosphorus entering the Mississippi (figure 8.2).[13]

Unfortunately, the increased pollution comes at a time when the river is ill-equipped to deal with it. For the past 200 years, humans have worked to straighten, modify, and reengineer the Mississippi, eliminating many of the natural mechanisms the river uses to defend itself. The Mississippi is the most dammed river on Earth, with 703 impediments.[14] To make the river navigable for large barges, the U.S. Army Corps of Engineers

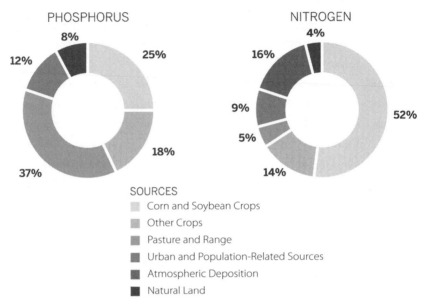

Figure 8.2. Fertilizers used to grow corn and soybeans in the Mississippi River Basin are the biggest source of nutrient pollution that ends up flowing into the Gulf of Mexico. *(Credit: The Freshwater Trust, adapted from U.S. Geological Survey data.)*

has dredged a 9-foot navigation channel on the river using "river training structures" such as chevrons, wing dikes, and bendway weirs.[15] And to carve out floodproof land for agriculture and urban areas, it has built levees that separate the river from its former floodplains. Today, the Mississippi is disconnected from 50 percent of its floodplain in the upper river and 90 percent in the middle and lower stretches.[16]

What's more, 80 percent of the original wetlands in midwestern states have been drained, much of it to make way for more ever-expanding agriculture.[17] In the eight states of the upper Mississippi River Basin, where 50 percent of all U.S. corn is grown, 35 million acres of wetlands—an area the size of Illinois—has been lost (figure 8.3).[18] As an Illinois native, I can tell you that's nothing to sniff at.

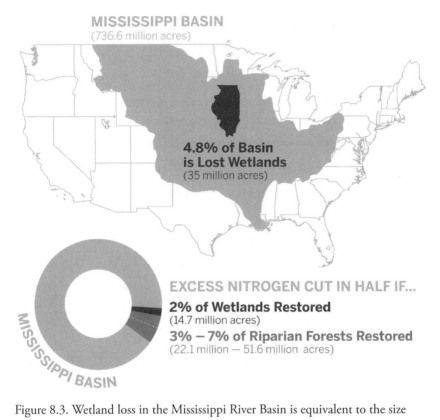

Figure 8.3. Wetland loss in the Mississippi River Basin is equivalent to the size of Illinois. Yet only a small and targeted portion of that area must be restored to significantly reduce the size of the dead zone at the Gulf of Mexico. *(Credit: The Freshwater Trust, using data adapted from Brooke Barton and Sarah Elizabeth Clark, "Water & Climate Risks Facing Corn Production: How Companies and Investors Can Cultivate Sustainability," a Ceres report, June 2014.)*

We now know these wetlands play a wide range of vital roles. Wetlands are essentially our nation's kidneys. They serve as a filter, soaking up nutrients and sediments before they flow into the river, purifying water and making it reusable. They also regulate the flow of the river by absorbing water, recharging aquifers, and allowing water to slowly seep into the river over time. They control flooding and erosion the natural way, by giving rivers more room to overflow. And they allow freshwater and sediments to spill over from the river, laying the foundation for a diverse ecosystem including birds, wildlife, and fish.

Unfortunately, the amount of nutrients flowing into the river isn't predicted to let up anytime soon. Both the amount of nutrients and the size of the dead zone at its mouth are expected to expand as a result of the federal mandate to grow corn for ethanol. Already, about 40 percent of all corn grown in the world comes from the Mississippi River watershed.[19] In 2007, the U.S. Congress set a goal of producing 36 billion gallons of ethanol a year by 2022. To do so, corn production will need to triple over 2006 production levels, much of it produced in the Mississippi River Basin. Expanding production to meet the ethanol mandate could by some estimates add as much as 19 percent of nitrogen pollution to the river's existing loads.[20]

The water–energy nexus tightens like a noose when you add in the effects of oil and gas drilling. The surge in hydraulic fracturing, or fracking, in places such as the Bakken, a 200,000-square-foot rock formation underlying parts of North Dakota and Montana, requires tremendous amounts of freshwater to keep the oil flowing. In 2012, for example, the Bakken oil industry used 5.5 billion gallons of water, more than all the residents of Fargo, North Dakota's largest city.[21] Something on the order of eight barrels of water to produce one barrel of oil is no joke.[22] Fracking wastewater can contain massive amounts of salt, toxic chemicals, and radioactivity that are difficult to treat and sometimes end up in the groundwater and nearby streams and rivers.[23] In the Bakken, fracking increases the risk of groundwater contamination along the floodplains of the Missouri River, which flows into the Mississippi. Although fracking in the region has only recently taken off, there are already 800 wells that directly place the health of the Missouri River and Lake Sakakawea at risk. In 2013 alone, more than 1,700 spills occurred due to fracking in the Bakken, a quarter of them uncontained as they've seeped into groundwater, wetlands, and streams.[24] As fracking runs through booms and busts in the coming decades, the risk to the Mississippi River Basin water quality will only continue to grow, partly from groundwater problems already put in motion by past action and partly from further injections.

Fracking in the Bakken also presents another risk to the Mississippi as more crude oil is carried by barge to oil refineries. In the 4 years from 2008 to 2012, for example, crude oil barge shipments in the United States have increased by 200 percent, much of it along the Mississippi River. As more oil is transported by barge, accidents are occurring. In 2011, for example, a barge crashed into a bridge, sending 11,000 gallons of crude oil into the river. And in 2014, a barge crash spilled 13,500 gallons of crude oil into the Mississippi, harming marine ecosystems and temporarily halting transportation on the river.[25]

All this increased pollution is coming at a time when climate change is putting greater pressure on the river. Over the last couple of years, shallow waters caused by some of the worst droughts on record have made parts of the Mississippi River impassable just as harvests were ready to be transported to market. Not only did these shallow waters put billions of dollars' worth of goods at risk of not reaching their destination, but there's been less water, and therefore less oxygen, to dissolve the nutrients, which concentrates the polluting effect.[26] At other times, heavy downpours have created massive floods in the basin, such as the 2011 flood, which caused $2.8 billion in damages and harmed more than 21,000 homes and 1.2 million acres of agricultural land.[27]

Taken together, these threats spell disaster for one of America's most important natural resources. It's no wonder why American Rivers has named the Mississippi one of America's most endangered rivers nine times since 1991.[28] As Loulan Pitre Sr., who has worked along the Gulf Coast his whole life, put it, "You can fool people. But you can't fool the fish."[29]

Repairing a Sinking Ship

Rivers speak. By their very nature, they show us that everything is connected. Every upstream economic action has its downstream ecological consequence, and vice versa. As our economic decisions add up to impaired rivers, we are gradually waking up to the link between a functioning environment and a prospering economy. When river volumes are too low, commerce suffers because full barges can't move.[30] When massive flooding occurs, businesses and agriculture suffer through building, infrastructure, and valuable soil loss.[31] And when we pollute our watersheds, the entire economy suffers, including business, the fishing industry, and tourism. When we weaken our rivers, both the economy and the environment sink to the bottom in the same leaky ship.

In recent years, many have attempted to repair that ship by allocating money and labor to a wide range of restoration efforts. Perhaps most notably, the U.S. government has spent billions of dollars on grants and landowner payments to reduce nutrient pollution. In 2009, for example, the U.S. Department of Agriculture introduced the Mississippi River Basin Healthy Watersheds Initiative, aimed at reducing the size of the dead zone. The initiative has allocated $320 million to help farmers implement management practices that reduce nitrogen and phosphorus from entering the Mississippi River.[32] This comes in addition to several other federal programs, such as the Conservation Reserve Enhancement Program (CREP) and the Environmental Quality Incentives Program (EQIP), which pay farmers and ranchers to remove environmentally sensitive land from production and implement less polluting farming practices.

Although these well-intended federal programs have certainly reduced some amount of nutrient pollution, they haven't allocated nearly enough funding to solve the problem. Throwing millions of dollars at a billions-of-dollars problem isn't going to make a significant dent in the long run. What's more, these efforts haven't embraced the principles of quantified conservation to ensure that even these dollars are put to the best use. In general, grants are awarded to any landowner that meets a broad set of qualifications—not necessarily the projects that capture the highest return for the environment.

Many states in the Mississippi River Basin have also adopted wetland mitigation banking programs. Federal laws such as the Clean Water Act and the Food Security Act require that damage to wetlands be mitigated. To comply with these regulations, some commercial developers, such as large shopping center developers, are offsetting their ecological disturbance by buying credits used to restore wetlands in other areas.

Yet here again, the program doesn't reflect the size of the problem, nor do the projects reliably render the proper environmental gain. To be effective, wetland restoration must happen on a much larger scale. Lost wetland acres in the basin, originally sited and designed by nature over time, are the size of Illinois. We started too late with a "no net loss" effort, and we need to improve.[33] Whether by mitigation banking or through direct compliance, the amount of wetland restoration along the Mississippi River Basin hasn't come close to what's needed to stop nitrogen and phosphorus from flowing into the river. According to one study, wetlands restoration needs to grow by a factor of ten to twenty-five times of what currently exists for it to have a serious impact.[34] What's more, many of the projects haven't been properly designed. In many cases, the sites chosen for wetland restoration are small pieces of land bordering shopping centers. These sites aren't areas where large amounts of nutrients would otherwise flow into the river. Nor are they realistically going to be used as habitat by birds and other wildlife. In other words, the credits aren't set properly, and the patchwork of new wetlands doesn't offset the initial damage generated by draining the original wetlands.

Any effort to restore the Mississippi is a step in the right direction—or at least a half-step. Yet so far these efforts have been no match for the severity of the problem. We are trying unsuccessfully to patch the leaks when what we need is a brand new ship, one that doesn't leak at all. If we are to restore the Mississippi River in a meaningful way, our outcomes must be far clearer and far bolder. Our work must be implemented on a far larger scale. Our solutions must attract the funding and involvement from every sector of society. And our efforts must be measured and monitored over time to make sure they're delivering the expected results that allow us to achieve the gains needed to address the problem.

A Quantified Approach to Restoring the Mississippi

As with any restoration effort, restoring the Mississippi starts with a thorough understanding of the current situation. Only by knowing how many excess nutrients are flowing into our rivers and how much the rivers can withstand can we take the actions needed to obtain real progress. We know the first of these answers, thanks to the efforts of the U.S. Geological Survey and the U.S. Department of Agriculture, which measure nutrient loads at various points along the Mississippi, monitoring how they change over time.

Yet we don't know what the pollution threshold is for the entire river. Under the federal Clean Water Act, states are required to develop total maximum daily loads (TMDLs) that define the amount of pollution an impaired waterway can withstand while still meeting the fishable–swimmable–drinkable water quality standards. But, because the Mississippi River flows through so many states, setting such limits on a state-by-state basis isn't enough. TMDLs were supposed to be conducted on all waterways by 1982. Yet today, it would be generous to say 60 percent have been completed nationally, with most of those being legally attackable on either process or substance. Worse than that, a study by the U.S. General Accounting Office (GAO) found that many of the TMDLs that are in place are vague, don't adequately identify what actions need to be taken, and aren't likely to attain water quality standards.[35] In a watershed such as the Mississippi River, where rivers and their pollutants travel across many state boundaries, we should have a single TMDL for the entire basin, designed to actually attain water quality standards rather than simply meet procedural standards.

When it comes to groundwater, we have even less information. The federal government doesn't consistently measure the health of our groundwater. Nor does it set thresholds for how much pollution our groundwater can withstand while still being safe to drink. State data are just as spotty if not more so.

Only through a solid real-time understanding of the situation can we set outcomes that allow both a healthy river and a healthy economy. Just as the European Union sets a total allowable catch for the number of fish that can be caught while still allowing the fish population to replace itself, we need to set pollution limits for the entire Mississippi. Once we've set these limits, we can then allocate percentages to each of the thirty-one states within the basin.

Some basin-wide goals have been set. In 1997, the U.S. Environmental Protection Agency (EPA) created a Hypoxia Task Force aimed at reducing the dead zone in the Gulf of Mexico. The task force set the goal of reducing the dead zone to 5,000 square kilometers by 2015 and created a second goal of reducing nitrogen and phosphorus loads across the basin by 45 percent. It also asked each state within the basin to implement a nutrient reduction strategy.[36]

Yet nearly 2 decades after the task force's formation, the dead zone hasn't shrunk, and none of the task force's goals have been met. Although the 2015 deadline is fast approaching, most states haven't even completed their nutrient reduction strategies, nor have they committed to specific reduction targets or timelines. The EPA's Office of Water has attributed the lack of progress to the cost of implementing these strategies, the unpopularity of doing so from various constituencies, and the fact that the EPA has not held these states accountable.[37] But inaction doesn't diminish resistance, and it certainly doesn't get us closer to success.

Recognizing things were on the wrong track, a group of environmental groups led by the Natural Resources Defense Council has sued the EPA to force the agency to use its authority under the Clean Water Act to enforce water quality standards where states have failed.[38] As a result of this suit, a federal district court ruled that the EPA needs to determine whether states are sufficiently solving the problem. If not, the agency will be required to propose its own standards, which could pave the way for federal numeric limits and stricter pollution controls.[39] Federal limits would give us a clear way to move forward while paving the way for the implementation of a water market across the entire basin.

Yet, even without federal limits, the good news is that it's possible to restore the Mississippi and to do so in a targeted way, one that doesn't involve every landowner but instead focuses on the most important places along the river. According to one analysis, for example, nitrogen levels could be nearly halved if just 2 percent of the Mississippi Basin were restored as wetlands and 3 percent to 7 percent of the basin were returned to riparian forest. With proper targeting of less than 10 percent of that basin, we would keep the economy moving while getting the environmental improvements we need, especially if we engage quantified conservation and the methods it enables.[40]

With today's data and analysis, we know that all restoration projects don't result in equal gains for the environment. For example, if just 7 percent of the Illinois River Basin watershed were converted to wetlands, as much as 50 percent of nitrogen from that watershed could be prevented from entering the river. By contrast, the James River Basin in South Dakota contributes just a small amount of nitrogen to the overall problem, so it doesn't make sense to target that river in the same way.[41]

Using existing technology, we can easily figure out which restoration projects will reap the largest returns for the environment and focus our efforts in those areas. Overlaying layers of data from geographic information systems (GIS) onto satellite photos on a Google Earth platform, for example, we can see precisely how different parcels of land are being used along the river down to the property-owner level.[42] And using software such as The Freshwater Trust's Basin Scout method, we can combine those satellite photos with physical data about the river, land, and cropping rotations and then use that information to assign quantifiable environmental gains that can be translated into credit values. If a farmer's growing dryland wheat, it's not a priority restoration project, and the credit values won't be high. But if he's growing corn that's heavy in fertilizer, the project would be assigned a high priority because we can get big environmental improvements if the farmer changes practices. Today's technology gives us a highly visual way to pinpoint the most important restoration sites with precision and speed—in effect, a Google search for the best opportunities for environmental gain. Once we've identified those sites, it's a matter of providing the right financial incentives to motivate farmers and livestock producers to implement less-polluting farming practices or shift ecologically sensitive land out of traditional agricultural production and into generating desperately needed bushels of nature.

Providing meaningful financial incentives requires creative funding mechanisms. It also entails the involvement of a wide range of interests: municipalities, factories, and power plants, government, farmers, businesses, environmentalists, and private investors. Within a quantified conservation framework, we can bring all these players together to restore the Mississippi River at a pace and scale that match the enormity of the problem.

Obtaining the Maximum Value for Every Taxpayer Dollar

Even as the targets get developed, we need to get smarter and get going on the ground. We spend a ton of federal and state money without understanding what it gets us. One change that could make a huge difference is the package of executive orders and legislative refinements (discussed in Chapter 2) that require the quantification of environmental improvements made with public investments. By retooling existing government programs, we can get a bigger restoration bang for every taxpayer dollar spent. And by embracing a quantified approach to conservation, both federal and state governments could make their existing funding programs do so much more for the Mississippi.

Just think if programs like the CREP and Mississippi River Basin Healthy Watersheds Initiative targeted only restoration projects that obtained the highest improvements for the environment. Just think if government monitored the success of its restoration projects to ensure that they were effective. And just think if it used these data to adjust its restoration programs over time to ensure that they were meeting their numerically defined goals. In essence, federal grants and landowner payments would be paying for quantifiable outcomes, not just restoration projects that may or may not be getting the desired result. If these simple changes were made, you can bet that the billions of dollars spent on restoring the Mississippi would achieve a lot more for the basin than they do today.

These are dollars that are already being spent. Quantified conservation simply ensures that they get spent in a smart way. By tracking the ecological improvement, we'll be able to tell a new story, one in which government and agriculture work together to make progress against a broad environmental goal.

Streamlining and Securing Their Supply Chains

In addition to government, the will to clean up the Mississippi must come from the business sector as well. Even without stricter regulations, this is already starting to happen because it makes financial sense. For example, one sector that's starting to advocate for a healthier Mississippi is the Gulf seafood industry, which supplies 40 percent of the nation's seafood.[43] The dead zone is costing the seafood industry $38 million a year, according to the National Oceanic and Atmospheric Administration.[44] Faced with lost revenues, the fishing industry has been joining forces with other groups fighting to eliminate the dead zone.

Moreover, multinational food companies from Coca-Cola to General Mills to Kellogg's are increasingly putting pressure on their supply chains to reduce their agrochemical and water use. One notable example: Walmart, the world's largest retailer, recently demanded that its food suppliers work with the U.S. farmers in its supply chain to improve the efficiency of fertilizer use by 30 percent on 10 million acres of corn, wheat, and soy by 2020.[45] These companies aren't taking these measures out of the goodness of their hearts but rather because they realize the supply of clean freshwater is limited, and they're attempting to reduce their supply chain risk. That business choice is paying an environmental dividend with less fertilizer headed downstream.

Getting More for Less with Water Markets

A basin-wide water trading program can be another big accelerator to cleaning up the Mississippi. For example, a recent pilot program conducted on behalf of the EPA by the World Resources Institute (WRI) found that interstate nutrient trading on the Mississippi River Basin is a viable way to meet the Hypoxia Task Force's goal of reducing nitrogen and phosphorus loads by 45 percent. The report concluded that nutrient trading could help shrink the Gulf dead zone, while providing a win–win situation for wastewater utilities and farmers.[46]

As discussed in Chapter 6, the Clean Water Act requires industrial plants, utilities, municipalities, and others that discharge water into rivers via pipes and sewers to treat and chill that water before returning it to the watershed. Historically, these regulated entities have done this by purchasing expensive water treatment plants or cooling towers that cost millions of dollars. Pretty spendy, with a mechanically limited benefit to the environment.

With a well-designed trading market, these regulated entities could instead offset their effluents by funding quantified restoration projects on private lands that motivate farmers and livestock producers elsewhere in the basin to implement environmentally friendly, best management farming practices or lease ecologically sensitive land to be restored. Trading gray infrastructure for green, landowners would be paid annually as long as the restoration project was maintained and monitored. Equivalencies matter, and these trades would happen in the form of standardized credits. Credits for nitrogen, phosphorus, sediments, dissolved oxygen, and temperature could be traded on the same market—in the same way that we use nickels, dimes, and quarters as trading currencies in our day-to-day lives.

A bedrock principle is that the credits must be designed so there's a net gain for the environment. Over the last couple of centuries, most of our decisions have achieved economic gain to the detriment of the river, and we need to restore the balance. So, for example, one unit of ecological disturbance could be counteracted by at least two units of environmental benefit in a place sited for success. Just as investors achieve financial gain when trading stocks, a well-designed trading program should create clear gains for the environment. What's more, the restoration projects must be monitored over time to make sure they're providing the predicted environmental gain. If they're not, both the restoration model and the accompanying credits must be adjusted. To facilitate this adjustment, the full cost of monitoring and maintenance must be baked into each project; although this need not be burdensome, we cannot skimp on data in a data-driven effort.

To generate the volume needed, trading should take place across the entire Mississippi River Basin but must have a proper structure. The units of trade must be standardized, tracked, and valued relationally—meaning there must be a predictable credit differential between where the restoration occurs and where the impact happens. Of course, nothing can be released into the river that knowingly reduces aquatic health under any circumstances. And ideally, if a wastewater treatment plant creates ecological damage by discharging a nutrient into a specific tributary of the Mississippi, the counteracting ecological benefit should happen in that same tributary.

The watershed still benefits if that wastewater treatment plant buys credits from a landowner further afield. Accordingly, the proper discounting must be applied, depending on where the trade takes place. For example, if a wastewater treatment plant buys a credit from a landowner within the same tributary, it would obtain full credit. But if it buys a credit from a landowner 500 miles upstream, it would receive only partial credit. Note that the landowner gets paid in full each time for the full bushels of nature his or her land produces. Weighted credits that take factors such as these into account are a vital component of a well-designed water market because they create more opportunity for trading to take place while catalyzing more work on the ground.

It's important for all the trades to be tracked and registered to ensure performance. What's more, there can be no double-dipping: If phosphorus credits are sold from a specific piece of property, you can't then sell temperature benefits from that same portion of property at a later date. This is a key way to get gain, right? We get more for the environment than just what is paid for.

For a conservationist, the central elegance of a water market is that it can achieve far more for the environment at a lesser cost than technological fixes. Gray cooling towers and filtration devices are expensive, and they just offset one parameter: They cool the water by a specific amount or filter out a specific pollutant, and that's it. Solving these problems the natural way accomplishes far more. A strip of trees doesn't just provide shade for the river. It stabilizes the bank. It sequesters carbon. It prevents nutrients and sediments from entering the watershed. It provides wildlife habitat. And it can act as a natural floodplain. It provides multiple benefits that aren't possible with engineered solutions.

A water trading market would also free up billions of dollars that are currently spent on solutions that don't add up to a lot of benefit for the environment. According to the WRI report, the participation of just two wastewater utilities in a Mississippi basin trading program would generate $715 million in restoration spending over a 20-year period. At the same time, wastewater utilities could cut their pollution reduction costs by $900 million, a 63 percent savings.[47] That's new money for conservation, with multiple benefits to everyone, from native nurseries to local restoration professionals to landowners growing bushels of nature to monitoring techs.

It All Comes Down to Design

As environmental markets have sprouted up, so too have their critics. Some attack market-based environmental solutions from an ideological perspective, arguing that it's not right to let industry pay to pollute. Others say we should be doing everything we can to eliminate river pollution, rather than polluting in one place and then offsetting it in another. And I get it. We used to advocate for many of the same things, until we realized our efforts simply weren't adding up. Zero pollution is a noble goal. But the reality is that we're never going to eliminate pollution altogether and still have a functioning economy. The trick is to figure out the right balance and then use data and analysis to make sure we're operating within our limits while gaining ground that's been lost over the last generation or two.

Others question the effectiveness of water markets, and often rightly so. Although many regions are starting to experiment with environmental markets, the majority have not been properly designed. Take the Chesapeake Bay water quality trading program, for example. Officials there have spent endless years and millions of dollars on process and planning, yet the results haven't added up to much. The problem is that the Chesapeake Bay stretches across six states and the District of Columbia, yet trading doesn't take place across the whole region. Instead, trading is restricted to state boundaries, with each state setting up trading credits according to its own rules. For example, some states allow sediment trading in addition to nitrogen and phosphorus, whereas others do not. Some states allow the participation of stormwater construction projects, whereas others do not. And the rules about buying and selling credits vary widely.[48] Without a single set of legal, biological, and transactional mechanics, the first deal isn't like the 5th, which isn't like the 40th. Such organized ad-hockery will never make it to the 4,000th deal. Although some states are doing good work pioneering these ideas, the bottom line is that there's no standardized market. In the meantime, pollution in the Chesapeake remains a huge problem.

Think of environmental markets as you would a rocket. If built correctly, a rocket can take us to wondrous places fast. But if it's not, it can tear across the landscape and destroy all manner of things. In the same way, markets can be brutally efficient, but they're only as effective as they are well designed. Just like a financial market, an environmental market needs to be built correctly. It needs to set the right pollution and water withdrawal limits. It needs to establish the right credit valuations to ensure a net gain for the environment. It needs to be standardized across the entire watershed. And, perhaps most importantly, the offsets need to be measured and monitored over a long period of time to ensure that they deliver the benefits as promised.

Demanding a Measurable Return on Their Investment

Efforts to restore the Mississippi River will be more effective as environmental groups, foundations, and private investors demand more from their financial investments. As the Gulf dead zone has attracted national attention, an increasing number of groups, foundations, and donations are focused on cleaning up the Mississippi River.

As more of these groups demand measurable returns on their investments, we will see more restoration for every dollar spent. I look forward to the day when these investors buy environmental outcomes in the form of quantified restoration projects that have already been implemented and verified by a third party. Rather than simply donating to a cause without knowing how the money will be spent, investors will purchase specific outcomes—such as 1 billion kilocalories of watershed cooling or a 10,000-pound reduction of nitrogen—on a key tributary. These purchases will be available through an online registry, making saving our rivers as simple as clicking on a specific restoration project at the landowner level, seeing the exact outcome that's been modeled, and then making the purchase over the Internet. Fully trackable and transparent.

Growing Bushels of Nature

Over time, I expect to see a new way of understanding the value of nature. As the environment becomes prized not only for what can be extracted but also for the services it provides to humans when it functions properly, the motivation to restore the Mississippi will also come from farmers, livestock producers, and landowners in the basin. As the financial incentives become more predictable, landowners will have the opportunity to participate in restoration programs and water trades that supplement their income.

Yet, even without these incentives, many farmers are realizing that it's in their own economic interest to implement conservation practices. Moving to precision farming gives farmers a competitive advantage as an increasing number of food companies demand it from their supply chains. What's more, these practices save farmers money. For example, farmers who participated in WRI's pilot program improved their financial situation by implementing conservation practices that reduced their total nitrogen and phosphorus use, even without water quality trading.

By implementing these practices, farmers both lowered fertilizer costs and increased crop yields. Interestingly, the study found that some of the main reasons these farmers had been reluctant to implement these conservation practices is because they didn't understand the opportunity or because they believed that using less fertilizer meant lower yields, and before, lower yields always meant less money.[49] But in agriculture, it's not always what you make but what you don't spend that creates profitability.

Although awareness hasn't made it across the board, things are slowly changing. Already, one third of all corn acres in the United States are farmed using best practices for nitrogen management, such as using less fertilizer and applying it at the right times.[50] Even if just a few farmers and livestock operators adopted the conservation practices of precision agriculture, they could have a large impact on the river. For example, if just 10 percent of farmers along the Mississippi substituted perennial crops such as alfalfa for some of their corn–soybean plantings, they could reduce nitrogen in the watershed by 0.5 million metric tons a year—one third of the total nitrogen that flows into the Gulf.[51] Similarly, if livestock producers engaged in better animal manure practices that reduced their runoff by just 20 percent, it would reduce nitrogen loads in the river by the same amount.[52] And that, in addition to all the other pieces we just covered, would go a long way toward improving the watershed.

Closing Our Open-Ended Accounting System

Restoring the Mississippi—as with any river—demands that we move beyond our open-ended ecological accounting system to one that really tells us where we stand and what we need to do next. Just as financial systems operate with a finite pot of money, with debits and credits, the earth is a closed-loop biosphere with limited natural resources. We need to acknowledge these limits, especially now, as they are coming into clear view.

My dad, the small-town carpenter, had a heavy-duty guarantee. Not written, but a guarantee that he stood by. I recall one particularly tall farmhouse he worked on that had an old chimney that kept leaking. Every summer for several years, we'd go out to the farm and patch that chimney until it was fixed. For me, I don't care what product you make, what service you provide, or what business you run. You need to stand by your product, and today, that includes being responsible for where it ends up at the end of its life cycle. This is how good businesses operate—and the only way they will operate in the future.

Businesses must pay for the ecological disruption that accompanies their product development, in a way that creates a net gain for the environment. I'm not a basher, but I do not shy from telling the truth. It's unconscionable that oil and gas outfits such as Halliburton are allowed to turn eight barrels of clean water into eight barrels of polluted water and walk away with a barrel of oil. For every barrel of water they pollute, they should return two barrels to the river or aquifer from which it came—by retiring water rights, for example, or converting farmers on the west side of the Missouri from irrigation to dryland cropping. In the same way, agrochemical companies such as Monsanto and PotashCorp shouldn't be allowed to simply manufacture fertilizers and turn a profit without being accountable for the accompanying downstream ecological damage. There are not that many fertilizer companies out there; a small percentage of their profits could form a fund to purchase and retire environmental credits generated through better farm management. Although this could significantly reduce their legal exposure to class action suits down the road driven by the presence of their chemicals in water supplies, I do not see this as punitive but rather as standing for one's product all the way through—and not leaving it for the next guy.

Government, too, needs to adopt a broader accounting system that takes the environment into full consideration and not work at cross-purposes. It's not enough to simply develop policies that increase demand for ethanol. The repercussions for water quality must be factored into the equation, and the potential damage either lessened or offset.

And the environmental community needs to step it up too. The time for pointing out problems without bringing a workable solution to the table is hereby over. We are in this together—way more than we realize.

The enormity of the Gulf dead zone is a testament to the sheer size of today's economic engine at work, unfortunately, much of it at the expense of the environment. Whether through disregard, denial, or indifference, we have collectively chosen a world in which economic impacts are dwarfing the needs of a healthy environment. We can restore the Mississippi River. We have both the technological skills and the scientific know-how to make it happen. Yet doing so will require us to rapidly move beyond today's inaccurate accounting system that fails to internalize the consequences of our decisions to a new construct that considers the wider implications. Achieving prosperity within the limits of our closed-loop biosphere is a paradigm shift. Yet I believe the perfect storm is brewing, and it's going to sweep up every sector of society in favor of a smarter approach.

CHAPTER NINE

It's Now and It's Us

A GENERATION AGO, an 18-year-old boy, Andy Lipkis, was so concerned with the smog that enshrouded Los Angeles that he founded a nonprofit focused on planting and preserving trees. His organization TreePeople began by raising $10,000 to plant 8,000 seedlings in the mountains surrounding Los Angeles. And seeing the valuable role that trees play in capturing and storing water, Lipkis eventually broadened his nonprofit's mission to include water conservation.

When Lipkis first started his organization, his ideas about using technology to replicate a tree's natural water storing abilities drew little more than raised eyebrows. And his proposals for making Los Angeles more flood resistant and less reliant on imported water were flat-out rejected.

Watershed management was for rural areas. Water was cheap. And officials saw no reason why they shouldn't continue importing water through the mountains from sources hundreds of miles away. "Twenty years ago, no one in government recognized the Los Angeles environment as a watershed," Lipkis said. "When I talked about it that way, people thought I was nuts."[1]

Forty years later, and Los Angeles's population has exploded, creating more demand for water. Climate change has led to persistent droughts, rendering the supply less reliable. And Los Angeles is struggling to provide enough clean water for its citizens. In short, the situation has changed, and Lipkis's ideas have gradually caught on.

In 1998, Lipkis proved that he could re-create a tree's natural rain capture process by staging a mock flood on a retrofitted Los Angeles house. He lined the house with gutters that channeled rainwater into cisterns. He lowered the front and backyard lawns to turn them into wetlands. He then brought in a water truck that dumped 15 tons of water on the roof. None of the water left the premises.[2] Watching the demonstration were flood control officials, who were so impressed that they abandoned a $42-million proposal to build a storm drain in a nearby flood-prone area.

Unlike "environmental groups that raise money for policy for various things" and aren't held accountable when they don't make a positive change, as Lipkis put it, TreePeople's on-the-ground work to capture and store rainwater clearly proved itself and is increasingly regarded as an important way to combat water scarcity.[3] So much so, in fact, that it may just change the future of water for Los Angeles. Already, the city has been installing underground water cisterns to capture and reuse natural rainwater rather than simply shunt it into a pipe out to sea. And in part because of Lipkis's demonstration project, the City of Los Angeles recently adopted a 20-year plan that treats the Los Angeles Basin as a single watershed that integrates water quality, water supply, flood control, and wastewater. Although the program is still in its infancy, Los Angeles already consumes less water than it did in 1970 despite the fact that its population has grown by more than one third (figure 9.1). And the city is on track to meet the ambitious goal of reducing its imported water by half within the next decade.[4]

If you want to know how a small number of people can move a big needle, Andy Lipkis and his organization are a powerful example. They saw a problem—a city that wasted its water—and decided not only to call it out but to fix it. They committed to waking up an entire citizenry to new possibilities and then managing Los Angeles's watershed in a holistic manner. They physically demonstrated how their solutions can work.

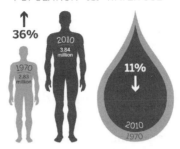

Figure 9.1. Thanks to Los Angeles's water conservation measures, the city consumes less water than it did in 1970, despite the fact that its population has grown by more than one third. *(Credit: The Freshwater Trust, using data from Jacques Leslie, "Los Angeles, City of Water," New York Times, December 6, 2014.)*

And they persevered even though their ideas weren't initially accepted. They could see where Los Angeles was headed, and they argued for a different path. As the water situation worsened for Los Angeles, their ideas gained acceptance. Not only are they changing how water is managed in Los Angeles, but their ideas are influencing urban water management throughout the country.

In a real sense, these ideas for water conservation were ahead of their time. When Lipkis first started arguing for better water management, importing water was cheap, and the will to conserve it just wasn't there. But now that we're bumping up against the limits of clean freshwater, interest in conserving water is growing, and creative, on-the-ground solutions like the ones TreePeople are proposing are getting traction.

The same thing has become true on a broader scale. As the limits of all of our natural resources come into clearer view, the time is ripe for new, creative solutions to conserve them. As environmentalists, we must move beyond the exclusive focus on advocacy and litigation to embrace on-the-ground solutions that improve our natural areas. The world is hungry for such solutions—not just because we're tired of the old ways of playing the game but because the old ways just don't get the job done anymore. Our future depends on shifting to strategies that actually work beyond paper and beyond our current era.

Debunking the Old Paradigm

In the 1960s scientist–historian Thomas Kuhn popularized the concept of a paradigm shift. Kuhn studied the difficulty of changing the minds of scientists, such as overcoming the assumption that the sun revolves around the earth. Such assumptions are broadly accepted by society, so much so that they're considered "the truth." Therefore, any attempt to introduce a new set of assumptions is resisted and even condemned. Think of Italian physicist and astronomer Galileo being convicted of heresy and forced to live under house arrest. His radical notion that the earth circled the sun didn't just conflict with the dominant social view of the time; it contradicted more than a thousand years of religious doctrine. The problem for the old-liners was that Galileo was simply observing reality; when they looked past their own blinding dogma, they saw for themselves.

Eventually we figured out that Earth is not the center of our universe and that our world is not flat. But have we debunked all that we must debunk? No. Looking at the environmental problems we face today, it's clear that we need a new paradigm. The original assumption underlying our current paradigm is that our natural resources are infinite and that we can use as many as we like without consequences. But as we run out of these resources and undercut the natural services they provide humans, a new observable reality is starting to contradict the existing arrangements and assumptions, causing them to fall into doubt. Indeed, in many parts of the world the current paradigm has already been thrown into crisis, and this will continue. Turns out the future is already here; it's just that it's not evenly distributed. Yet.

For evidence that a new way of looking at the world has been taking hold, check out almost any website that discusses ecological issues, and you'll read about nature performing vital services that can be assigned a quantified value and translated into dollars. The idea of ecosystem services has been rapidly spreading, and it's gradually replacing the old idea that the economy and the environment are at opposite corners of the boxing ring. The idea that what's good for the environment is bad for the economy has passed its "use-by" date, and it's giving way to a new paradigm in which the economy and the environment are an integrated whole. Tradeoffs are known and participants are accountable. The central tenet is that a healthy environment is essential to a thriving economy and vice versa. And the result will be that gains for the environment will be viewed alongside gains for the financial sector.

Giving the Environment Legitimacy

A shift of paradigm, by definition, entails a change of large magnitude. And for that to happen, people need to adapt their thinking, as the public eventually did in how they thought about our planetary system. A paradigm is also community based. It's the assumptions around which society organizes itself. So changing the paradigm involves a shift in human consciousness. As American social scientist Willis Harman wrote in his book *Global Mind Change*, "Throughout history, the really fundamental changes in societies have come about not from the dictates of governments and the results of battles but through vast numbers of people changing their minds—sometimes only by a little."[5] People give legitimacy to an idea, and they can take it away, Willis said, just as Americans did when they terminated slavery as an acceptable institution.

As the effects of mismanaged water health reach into the home—as they have from Charleston, West Virginia, to Toledo, Ohio, to California's Central Valley and many other places around the country—people are changing their minds, and they're legitimizing the idea that the environment has a value. At the same time, they're challenging the notion that a growing economy is all that matters. Yes, jobs are important, and we need them to survive. But yes, clean water is also important, and we need that for our survival too. People are seeing that a bankrupt environment means a bankrupt economy. And bit by bit, we are seeing that a properly designed economy can catalyze real environmental strides. Although many argue whether the economy or the environment is the greatest or most important force on Earth, I see them as two parts of a system that simply needs to evolve into a unified whole.

Consider how a caterpillar turns into a butterfly. As the metamorphosis occurs, highly organized groups of cells called imaginal discs start to show up in the body of the caterpillar, and most are wiped out because they're not recognized by the caterpillar's immune system. But they keep at it, and as more arise, they begin to overwhelm the caterpillar's immune system. In the end, the caterpillar's body deteriorates and the imaginal discs build the butterfly from the decayed materials of the caterpillar.[6]

In the same way, a new way of viewing our global economy is starting to overwhelm the old. You can see it seeping into the food system, which is starting to embrace precision agriculture and other conservation techniques that secure farmers' prosperity over time. You can see it seeping into business, which is starting to manage its water and agrochemical use across the entire supply chain. And you can see it seeping into the mindset of consumers, as they start to demand products that are produced sustainably and have a limited impact on the planet. Although the economy is still understood as today's dominant force, we are gradually adopting a new lens with which to view what it means to be prosperous on planet Earth.

The modern environmental movement of the 1970s began to shift our thinking by assigning the environment a value. Now we're completing the shift with the understanding that the environment and the economy must be integrated to be maintained. And as we change our paradigm, our desired outcome is changing along with it. Rather than maximizing economic returns without regard for the planet, we're developing a new goal, one that's focused on *optimizing* economic returns within the constraints of a healthy environment.

The good news is that, once we collectively establish this new outcome, we have the tools at our disposal to achieve it. From satellite photos to on-the-ground measurement tools to big data software frameworks such as Hadoop that can analyze vast amounts of data, twenty-first-century technology can help us assess the current state of the environment. We can now offset environmental damage in a precise, measurable way that creates a net gain for the environment. We can develop environmental markets and other quantified approaches that build in ongoing mechanisms for protecting the environment. We can innovate funding models that bring greater amounts of money to bear on our problems. And we can accurately measure whether our efforts are working, quickly making the necessary adjustments.

The 80/20 Principle

Just as in the days of Galileo, when a change of large magnitude happens, resistance to change is a given and many people dig in hard to maintain the status quo. Often, those who've most benefited from the old paradigm are the least willing to change—and this has implications for the environmental old guard as well as entrenched business interests. As Andy Grove wrote in his book *Only the Paranoid Survive*, "The person who is the star of the previous era is often the last one to adapt to change, the last one to yield to logic of a strategic inflection point and tends to fall harder than most."[7]

In the same way that the caterpillar's immune system initially resists its transformation into a butterfly, our brains are hard-wired to protect us, and part of that protection involves following the path that's already familiar and has been free of danger in the past. Sticking to what we already know is a great way to avoid discomfort.

Fortunately, creative thinking and the innovations that result have a way of building on one another. In fact, innovations happen far more often as refinements of an existing idea rather than as a brand new one. It's largely a honing process that makes good stuff work better, and the environmental pioneers should take pride in what their effort wrought while letting us push forward into a new world. As Walter Isaacson pointed out in his excellent book *The Innovators*, although the digital age may seem revolutionary, in reality it came about by expanding the ideas handed down from previous generations. "The best innovators were those who understood the trajectory of technological change and took the baton from innovators who preceded them," he wrote.[8]

What's more, change doesn't demand that everyone get in line. In business, there's an "80/20 principle" that says 80 percent of the results stem from 20 percent of the effort. This rule is often used to explain how sales occur—for example, that 80 percent of sales are typically generated by 20 percent of clients.

When it comes to change, the 80/20 principle applies as well. Changing a paradigm doesn't require the participation of everyone. Instead, 20 percent of the people can create 80 percent of the change. In other words, a small percentage of forward-thinking people have the power to create large societal shifts. The 20 just needs the conviction to proceed past old thinking.

Take the company Seventh Generation, for example. True to its name, the company manufactures and distributes environmentally friendly cleaning products while also weighing the impact of its decisions on the next seven generations. Seventh Generation was the first home care products company to remove phosphorus from automatic dishwashing products. It was the first to take a stand against volatile organic compounds (VOCs). And it became the first to voluntarily disclose ingredients on its product labels. None of these things were required. Yet Seventh Generation took the lead on its own, building a thriving business within the parameters of a healthy environment. "We started Seventh Generation with truth and purpose," said the company's leader, John Replogle. "Everything we have done over the past 26 years has been dedicated to the notion that there is a better way."

Interestingly, Seventh Generation's conservation practices are disrupting the entire home care product industry. As the company has attracted a loyal and growing customer base, large competitors from Clorox to Procter & Gamble to SC Johnson have taken notice, and they've responded by introducing their own sustainable products. "Isn't that what we should all be looking for—the opportunity to disrupt?" Replogle told me. "That is the truest form of innovation." Seventh Generation's approach to change is an example of the 80/20 principle at work. Great ideas can be contagious, and it often takes just a few people to set the wheels in motion.

The Misfits and the Rebels

American radio host Thom Hartmann is known for his thought-pro-voking broadcasts. And in one of his shows, he talked about a group of researchers who were studying a troop of chimpanzees in Kenya. The researchers identified the depressed chimpanzees, the ones who stayed on the perimeter of the group, and they tranquilized these chimpanzees and pulled them out of the area to see what would happen.

With the nonconformists gone, the researchers expected the remaining chimpanzees to turn into party animals. But a year later, the entire troop of chimpanzees was dead. Why? Because the ones on the perimeter served as the early warning system. They were hypervigilant—the ones who noticed the python making its way through the jungle, or the lion trying to sneak up on the group. During times of danger, they were the ones who sounded the alarm bells.[9]

In the same way, those of us who see the coming crisis, who are sounding the alarm and pushing to implement new solutions, have a critical role to play. The "system," if you will, is designed to make people get in line and follow the rules. And if we don't, it's easy to feel discouraged and sometimes even ostracized. Yet the truth of the matter is that those of us who are willing to step out of line and work toward a new paradigm are important to human survival. In fact, we need a lot more people who are willing to do so. As Steve Jobs so famously said, it's the "crazy ones, the misfits, the rebels, the troublemakers, the round pegs in the square holes . . . the ones who see things differently" who end up changing the world. Indeed, at a time when the old way isn't working, being crazy enough to try something new is exactly what the planet needs.

Remaking Our Organizations and Ourselves

In a world that's constantly changing, rigidity is a death sentence. At the business level, much has been written about the need to anticipate and adapt to change. In their seminal book *Reengineering the Corporation*, for example, authors Michael Hammer and James Champy urge businesses to move beyond corporate structures created during the Industrial Revolution to meet the new business challenges of the twenty-first century. "Change has become both pervasive and persistent," they wrote. "It *is* normality."[10]

Likewise, Peter Senge, in his famous book *The Fifth Discipline*, introduced the concept of the "learning organization"—that organizations need to continually foster learning as a way to expand their capacity to create the results they desire. "Through learning we re-create ourselves," Senge wrote. "Through learning we become able to do something we never were able to do. Through learning we re-perceive the world and our relationship to it."[11]

Indeed, being a lifelong learner and an organization that readily adapts to change is a top priority for any successful company. And whether we're in the business world or not, we all need to adopt that same ethic: Accept change as a fact of life and be willing to adapt. In chaos there is opportunity if we can keep our wits about us. Like the learning organization Senge writes about, we must continually re-create ourselves and the organizations we work for—and that entails being nosebleed honest with ourselves about when it's time to change.

At The Freshwater Trust, a close examination of our results revealed that the time had come. Had we continued down the same old path of advocacy and litigation, the end of the road would have been our own irrelevance. We probably would have hung on as so many do now, exhausted, yet supported by a world that neglects to honestly examine results.

I often tell people, "We're stubborn on outcomes but flexible on tactics." In other words, we're stubborn about getting results for our streams and rivers but flexible as to how we get there. The problem we are trying to solve affords us no time to care about conventional wisdom, especially when it just slows progress. Implementing the best tactics requires constant learning and adaptation. It also requires facing tough roadblocks head on and devising ways to remove or get around them. And if we're not getting around a roadblock, we need to stop and figure out why, but just long enough to understand, adjust, and try again; studying a problem too long creates inertia. As I see it, the sooner we make a mistake, the sooner we can find a solution and move on to the next one.

As the new paradigm takes hold, it's time for *all of us* to change or run the risk of irrelevance. We must move beyond the 1980s and recalibrate to bring our methods in line with the twenty-first-century problems we face. And we must be honest with ourselves about our impacts, both bad and good:

- If you're in *agriculture*, know your impact on the environment and work to reduce it. We can't privatize wealth and socialize risk. If you're growing corn, soybeans, or any crop, consider the fertilizer that's left on your field and manage it in such a way that it doesn't flow into the watershed. It will save you money.

- If you're in *manufacturing*, understand clearly what goes into your product and what happens to it at the end of its shelf life. Then re-design it in such a way that it has zero net impact. It will both lean up and secure your supply chain.

- If you're a *food company*, demand that the ingredients you purchase are grown in a sustainable way. Better yet, work proactively with your suppliers to change their practices. Consumers will demand it soon enough.

- If you're a *philanthropist*, it's time to require real results. Stop believing in the story of good effort that may someday add up and start relying on quantitative outcomes to ensure every dollar counts. Your mission impact will grow.

- If you're in the *environmental* or *government* world, it's time to risk trying new solutions that build toward an outcome rather than grind to a halt. Test your ideas to make sure they work, push to implement them broadly, and use data and analysis to make sure you're getting results. You will get the results you hoped for when you got into this work.

Perhaps most importantly, we all need to work together in a spirit of cooperation. It's time to end the finger pointing and start the thumb raising. Because, in the end, we're all in this together.

We Can Do This

As someone who is on the young end of the old and the old end of the young, I am attempting to translate the past into the future. When I talk to baby boomers about the concepts in this book, the reaction I sometimes get is that they are outlandish, even radical. But when I talk to millennials, they say, of course we should embrace new solutions. Of course we should harness twenty-first-century technology. And of course we should quantify our results. For most, quantified conservation is a no-brainer.

For millennials who grew up in a digital world, rapid, technological progress is a normal thing. And today's technology has created a generation of informed people who are educated about the environment and ready to act. Most millennials consider making a difference in the world to be more important than financial success.[12] They seek purpose, they want to create value and have impact, and they embrace a spirit of optimism, cooperation, self-discipline, and achievement.[13] With 78 million millennials in the United States alone, this itself is reason for optimism. Harnessing their energy and talent will bring about this quantified paradigm and a new era marked by environmental *gain*.

The truth is we can do this. And the bottom line is we must. No matter what our age or role in society, a better approach is here. With quantified conservation, we now have the tools to properly integrate the economy and the environment. It's time we leave behind the old choice of one over the other and hard pivot to a prosperity that combines both.

As James Jannard, founder of the innovative eyewear company Oakley, Inc., so pointedly noted, "Everything in the world can and will be made better. The only questions are, 'when and by whom?'"[14]

To this challenge, I answer: "Now, and by us."

Notes

Introduction

1. U.S. Environmental Protection Agency, Region 7, "Blackbird Creek, Adair County, Missouri: Total Maximum Daily Load, June 2006," http://www.epa .gov/region7/water/pdf/blackbird_creek_finaltmdl062706.pdf.

Chapter 1

1. U.S. Environmental Protection Agency, "Pesticide News Story: EPA Releases Report Containing Latest Estimates of Pesticide Use in the United States," February 17, 2011, http://epa.gov/oppfead1/cb/csb_page/updates/2011/sales"-usage06-07.html.
2. United Nations Development Program, "Issues Brief: Ocean Hypoxia—'Dead Zones,'" May 15, 2013, http://www.undp.org/content/undp/en/home/library page/environment-energy/water_governance/ocean_and_coastalareagover nance/issue-brief—-ocean-hypoxia—dead-zones-/.
3. Pacific Rivers Council, "Freshwater Ecosystems in Crisis: A Synopsis of Decline," Sec. 1:6, 2007, http://pacificrivers.org/science-research/resources-pub lications/freshwater-ecosystems-in-crisis.
4. Jennifer Horton, "What's Depleting Salmon Populations?" HowStuffWorks. com, November 18, 2008, http://adventure.howstuffworks.com/outdoor-ac tivities/fishing/fish-conservation/fish-populations/salmon-population.htm.
5. U.S. Fish and Wildlife Service, "Status and Trends of Wetlands and Deepwater Habitats," http://www.fws.gov/wetlands/Documents/Status-and-Trends -of-Wetlands-and-Deepwater-Habitats-in-the-Conterminous-United-States- 1950s-to-1970s.pdf and http://www.fws.gov/wetlands/Documents/Status-and

-Trends-of-Wetlands-in-the-Conterminous-United-States-2004-to-2009.pdf.

6. U.S. Department of Agriculture, "Major Uses of Land in the United States," 2002 and 2007, http://www.ers.usda.gov/publications/eib-economic-informa tion-bulletin/eib14.aspx and http://www.ers.usda.gov/publications/eib-eco nomic-information-bulletin/eib89.aspx.

7. Intergovernmental Panel on Climate Change, "IPCC Fourth Assessment Report: Climate Change 2007," http://www.ipcc.ch/publications_and_data/ar4 /syr/en/spms2.html.

8. Brigid Fitzgerald Reading, "Global Economy Expanded More Slowly than Expected in 2011," Earth Policy Institute, February 16, 2012, http://www.earth -policy.org/indicators/C53/economy_2012.

9. Alyson Shontell, "The Amazing Story of How Steve Jobs Took Apple from Near Bankruptcy to Billions in 13 Years," *Business Insider*, January 19, 2011, http:// www.businessinsider.com/how-steve-jobs-took-apple-from-near-bankruptcy -to-billions-in-13-years-2011-1?op=1k.

10. Steven Levy, "Google's Larry Page on Why Moonshots Matter," *Wired*, January 17, 2013, http://www.wired.com/2013/01/ff-qa-larry-page/.

11. Andrea James, "Amazon.com CEO Jeff Bezos Says Company Goals Not Changed," *Seattle Post-Intelligencer*, April 20, 2009, http://blog.seattlepi.com /amazon/2009/04/20/amazon-com-ceo-jeff-bezos-says-company-goals-not -changed/.

12. Brad Reed, "Amazon's Annual Kindle Sales Estimated at $4.5 Billion," BGR. com, August 12, 2013, http://bgr.com/2013/08/12/amazon-kindle-annual -sales-estimate/.

13. Jim Bubulsky, "Chaotic Storage Lessons: Why Drone Delivery Service Wasn't Amazon's Most Impressive Feat Highlighted on 60 Minutes," *Tech Talk*, December 11, 2013, https://medium.com/tech-talk/e3b7de266476.

14. J. Bennett, "A Look Inside Amazon's 'Chaotic' Warehouses," *Tech Spotters*, December 17, 2012, http://www.techspotters.co.uk/blog/a-look-inside-amazons -chaotic-warehouses/.

15. "Amazon Tests Drones for Same-Day Parcel Delivery, Bezos Says," *Bloomberg News*, December 2, 2013, http://www.bloomberg.com/news/2013-12-02/ama zon-testing-octocopters-for-delivery-ceo-tells-60-minutes-.html.

16. C. Kettmann, "The Rise of Big Data and Big Analytics, and the Inception of Return on Investment," *ViralHeat*, October 18, 2012, https://www.viralheat .com/blog/2012/10/18/big-data-and-big-analytics/.

17. Kokogiak Media, "One Quintillion Pennies," http://www.kokogiak.com/ megapenny/eighteen.asp.

18. Alan Horton, "Restoration Navigation: Charting a New Course for Conservation Investment," *Freshwater* magazine, Spring 2012, p. 17, The Freshwater Trust, http://www.thefreshwatertrust.org/freshwater-spring-2012/.

19. Brian Proffitt, "Peta, Exa, Yotta and Beyond: Big Data Reaches Cosmic Proportions (Infographic)," ReadWrite.com, November 23, 2012, http://readwrite .com/2012/11/23/peta-exa-yotta-and-beyond-big-data-reaches-cosmic-pro portions-infographic#awesm=~oCnZTv0yeYiQlY.

20. Growing Blue, "Smarter Water Management in Sonoma County, March 14, 2013, http://growingblue.com/case-studies/smarter-water-management-in-sono ma-county/.

21. IBM Website, "IBM Aims to Help Alleviate Water Shortages in Northern California's Wine Country," June 25, 2010, http://www-03.ibm.com/press/us/en /pressrelease/31995.wss.

22. Yale University Environmental Performance Index Website, "A Global Ranking for the Environment," http://epi.yale.edu/.

23. Organisation for Economic Co-operation and Development, "Moving Towards a Common Approach on Green Growth Indicators," April 2012, http:// www.oecd.org/greengrowth/greengrowthindicators.htm.

24. U.S. Department of Agriculture, "FY 2015 Budget Summary and Annual Performance Plan," p. 61, http://www.obpa.usda.gov/budsum/FY15budsum .pdf.

Chapter 2

1. U.S. Environmental Protection Agency, "EPA Approves Historic Salmon Restoration Plan for Klamath River, January 4, 2011, http://yosemite.epa.gov/opa /admpress.nsf/0/2BFCDE6032C3B3638525780E005BE8E3.

2. Jacques Leslie, "Oregon's Klamath River Basin One Step Closer to Historic Dam Removal," *Earth Island Journal*, April 17, 2014, http://www.earthisland.org /journal/index.php/elist/eListRead/oregons_klamath_river_basin_one_step _closer_to_historic_dam_removal/.

3. U.S. Environmental Protection Agency, "Chesapeake Bay," http://www2 .epa.gov/nutrient-policy-data/chesapeake-bay.

4. Chesapeake Bay Program, "Why Is the Chesapeake Bay So Important?" October 22, 2010, http://www.chesapeakebay.net/blog/post/why_is_the_ches apeake_bay_so_important.

5. Chesapeake Bay Foundation, "Nitrogen and Phosphorus," http://www.cbf .org/about-the-bay/issues/dead-zones/nitrogen-phosphorus.

6. James Price, Chesapeake Bay Ecological Foundation, Inc., "Chesapeake Bay

an Undeclared Ecological Disaster," July 22, 2014, http://www.chesbay.org
/articles/3.asp.

7. Chesapeake Bay Foundation, "The Economic Argument for Cleaning Up the
    Chesapeake Bay and Its Rivers," May 2012, p. 12, http://www.cbf.org/Docu
    ment.Doc?id=591.

8. Richard Heinberg, *The End of Growth: Adapting to Our New Economic Reality*
    (Gabriola Island, BC: New Society Publishers, 2011), p. 7.

9. National Public Radio, "The Vision of John Wesley Powell," August 26, 2003,
    http://www.npr.org/programs/atc/features/2003/aug/water/part1.html.

10. Ben Jervey, "John Wesley Powell's Watershed States Map: What If Our Western
    States Were Shaped by Watersheds?" *Good Magazine*, November 21, 2010,
    http://magazine.good.is/articles/john-wesley-powell-s-watershed-states-map.

11. Ibid.

12. Robert Glennon, *Unquenchable: America's Water Crisis and What to Do about It*
    (Washington, DC: Beacon Press, 2009), p. 304.

13. Ibid., p. 305.

14. Leonard F. Konikow, U.S. Geological Survey, "Groundwater Depletion in
    the United States (1900–2008)," 2013, http://pubs.usgs.gov/sir/2013/5079
    /SIR2013-5079.pdf.

15. Bridget R. Scanlon, Claudia C. Faunt, Laurent Longeuvergne, et al., National
    Academy of Sciences, "Groundwater Depletion and Sustainability of Irrigation
    in the US High Plains and Central Valley," March 14, 2012, http://www.pnas
    .org/content/early/2012/05/24/1200311109.abstract?sid=a0454585-baea
    -48aa-a1fb-96732c979d40.

16. National Ground Water Association, "Groundwater U.S.A.," http://www.ngwa
    .org/Events-Education/awareness/Documents/usfactsheet.pdf.

17. U.S. Environmental Protection Agency, "FACTOIDS: Drinking Water and
    Groundwater Statistics for 2009," http://www.epa.gov/ogwdw/databases/pdfs
    /data_factoids_2009.pdf.

18. Dave Owen, "Taking Groundwater," Washington University Law Review, Feb-
    ruary 14, 2013, p. 3. http://papers.ssrn.com/sol3/papers.cfm?abstract_id=22
    17770.

19. Juliet Christina-Smith and Peter H. Gleick, *A Twenty-First Century U.S. Water
    Policy* (Oxford: Oxford University Press, 2012), p. 14.

20. James Salzman, Slate.com, "Why Rivers No Longer Burn," December
    10, 2012, http://www.slate.com/articles/health_and_science/science/2012/12
    /clean_water_act_40th_anniversary_the_greatest_success_in_environmen
    tal_law.html.

21. Jan G. Laitos and Heidi Ruckriegle, "The Clean Water Act and the Challenge of Agricultural Pollution," *Vermont Law Review*, 37, no. 1033, p. 1045, http://lawreview.vermontlaw.edu/files/2013/08/14-Laitos-Ruckriegle.pdf.

22. U.S. Environmental Protection Agency, "Agricultural Nonpoint Source Fact Sheet," March 2005, http://water.epa.gov/polwaste/nps/agriculture_facts.cfm.

23. U.S. Environmental Protection Agency, "Nonpoint Source Pollution: The Nation's Largest Water Quality Problem," http://water.epa.gov/polwaste/nps/outreach/point1.cfm.

24. Ibid.

25. National Ground-Water Monitoring Network, "Establishing a Collaborative National Ground-Water Monitoring Network Program for the United States," November 2013, http://acwi.gov/sogw/NGWMN_InfoSheet_final.pdf.

26. Don Elder, Gayle Killam, and Paul Koberstein, *The Clean Water Act: An Owner's Manual* (Portland, OR: River Network, 1999), p. 82.

27. The White House Office of the Press Secretary, "Executive Order: Making Open and Machine Readable the New Default for Government Information," May 9, 2013, http://www.whitehouse.gov/the-press-office/2013/05/09/executive-order-making-open-and-machine-readable-new-default-government-.

28. The White House Office of the Secretary, "Memorandum for the Heads of Executive Departments and Agencies," March 9, 2009, http://www.whitehouse.gov/the-press-office/memorandum-heads-executive-departments-and-agencies-3-9-09.

29. The White House Office of the Press Secretary, "Executive Order: Identifying and Reducing Regulatory Burdens," May 10, 2012, http://www.whitehouse.gov/the-press-office/2012/05/10/executive-order-identifying-and-reducing-regulatory-burdens.

30. "Governor Brown Issues Executive Order to Redouble State Drought Actions," CA.gov, April 25, 2014, http://gov.ca.gov/news.php?id=18496.

Chapter 3

1. Brief for Oregon Trout as Amici Curiae Supporting Respondents, *Alsea Valley Alliance v. Evans*, 161 F. Supp. 2d 1154 (D.Or. 2001) (No. 99-06265).

2. *Alsea Valley Alliance v. Department of Commerce NMFS*, http://caselaw.findlaw.com/us-9th-circuit/1241890.html.

3. Effective May 12, 2008, NMFS relisted Oregon Coast ESU coho salmon. https://www.federalregister.gov/articles/2011/04/13/2011-8822/endangered-and-threatened-wildlife-and-plants-44-marine-and-anadromous-taxa-adding-10-taxa-delisting.

4. Philip Shabecoff, *A Fierce Green Fire: The American Environmental Movement* (Washington, DC: Island Press, 2003), p. 106.

5. "Earth Day: The History of a Movement," http://www.earthday.org/earth-day -history-movement.

6. Ibid.

7. Such legislation includes the National Environmental Policy Act of 1969, 42 U.S.C. § 4321; Clean Air Act of 1970, 42 U.S.C. § 7401; the Clean Water Act of 1972, 33 U.S.C. § 1251; the Coastal Zone Management Act of 1972, 16 U.S.C. § 1451; and the Endangered Species Act of 1973, 16 U.S.C. § 1531.

8. Shabecoff, *A Fierce Green Fire*, p. 131.

9. See *TVA v. Hill*, 437 U.S. 153 (1978); *Calvert Cliffs' Coordinating Committee, Inc. v. U.S. Atomic Energy Commission*, 449 F.2d 1109 (D.D.C. 1971); *NRDC v. Morton*, 458 F.2d 827 (D.D.C. 1972); *NRDC v. Costle*, 568 F.2d 1369 (D.C. App. 1977); *U.S. v. Earth Sciences, Inc.*, 599 F.2d 368 (10th Cir. 1979); and *Apex Oil v. U.S.*, 530 F.2d 1291 (8th Cir. 1976).

10. Robert Gottlieb, *Forcing the Spring: The Transformation of the American Environmental Movement* (Washington, DC: Island Press, 2005), p. 184.

11. David Helvarg, "'Wise Use' in the White House," Sierra Club Elections, 2004, http://vault.sierraclub.org/sierra/200409/wiseuse.asp.

12. Christopher J. Bosso and Deborah Lynn Guber, "Maintaining Presence: Environmental Advocacy and the Permanent Campaign," in *Environmental Policy: New Directions for the 21st Century*, 6th ed., Norman J. Vig and Michael E. Kraft, eds. (Washington, DC: CQ Press, 2005), p. 90.

13. Bosso and Guber, "Maintaining Presence," p. 93.

14. Ibid., p. 94.

15. Sarah L. Pettijohn, "The Non-Profit Sector in Brief: Public Charities, Giving, and Volunteering, 2013," Urban Institute, p. 4, http://www.urban.org/Up loadedPDF/412923-The-Nonprofit-Sector-in-Brief.pdf.

16. Russell J. Dalton, "The Greening of the Globe? Cross-National Levels of Environmental Group Membership," *Environmental Politics* 14, No. 4 (August 2005), p. 444, http://www.socsci.uci.edu/~rdalton/archive/green.globe.pdf.

17. Baird Straughan and Tom Pollack, "The Broader Movement: Nonprofit Environmental and Conservation Organizations, 1989–2005," National Center for Charitable Statistics at the Urban Institute, p. 25, http://www.urban.org /UploadedPDF/411797_environmental_conservation_organizations.pdf.

18. Pettijohn, "The Non-Profit Sector in Brief," p. 4.

19. John Roach, "Earth Day Facts: When It Is, How It Began, What to Do," *National Geographic News*, April 20, 2012, http://news.nationalgeographic

.com/news/2012/04/120420-earth-day-facts-2012-environment-science-na
tion/.

20. Union of Concerned Scientists, "1992 World Scientists' Warning to Humanity,"
http://www.ucsusa.org/about/1992-world-scientists.html#.VC8ADxbzZuB.

21. A 2013 U.S. Government Accountability Office (GAO) report on total maximum daily load (TMDL) plans developed under the U.S. Clean Water Act
(CWA) found the effectiveness of such plans lacking. With 50,000 TMDLs
developed since the CWA's inception, the GAO surveyed state TMDL managers and analyzed a sample of 25 TMDLs. It found that 83 percent of TMDLs
were achieving point source reductions, but only 20 percent of TMDLs had
met nonpoint source reductions. State coordinators reported that in 35 percent of NPS-only TMDLs, they did not know whether pollution levels had
changed. "Clean Water Act: Changes Needed if Key EPA Program Is to Help
Fulfill the Nation's Water Quality Goals," U.S. Government Accountability
Office, December 2013, http://www.gao.gov/products/GAO-14-80.

22. Michael Shellenberger and Ted Norhaus, "The Death of Environmentalism:
Global Warming Politics in a Post-Environmental World," p. 12, http://www
.thebreakthrough.org/images/Death_of_Environmentalism.pdf.

23. americanrivers.org.

24. columbiariverkeeper.org.

25. Joe Whitworth, "Holding the Line Means Losing," *Stanford Social Innovation
Review*, July 24, 2012, http://www.ssireview.org/blog/entry/holding_the_line
_means_losing.

26. Tom Knudson, "Environment Inc.," *The Sacramento Bee*, April 22, 2001,
http://journeytoforever.org/bflpics/EnvironmentInc.pdf.

27. Ibid.

28. Michaela Hass, "Cure for Environmental Fatigue: A Conference on Pathways
to 100% Renewable Energy," *Huffington Post*, March 14, 2013, http://www
.huffingtonpost.com/michaela-haas/a-cure-for-environmental-_b_2870788
.html.

29. Gottlieb, *Forcing the Spring*, p. 182.

30. Knudson, "Environment Inc."

31. Ibid.

32. Adam Werbach, "Where the Environmental Movement Can and Should Go
from Here," December 8, 2004, Grist reprint, http://grist.org/article/werbach
-reprint/.

33. Examples include *Citizens to Preserve Overton Park, Inc. v. Volpe* and *Chevron,
U.S.A., Inc. v. Natural Resources Defense Council, Inc.*

34. Knudson, "Environment Inc."

35. Ibid.

36. U.S. Chamber of Commerce, "Top 10 Environmental Myths," 2004, p. 4, https://www.uschamber.com/sites/default/files/legacy/reports/top10myths .pdf.

37. http://earthfirstjournal.org/about/.

38. Danielle Sacks, "Working with the Enemy," *Fast Company*, September 1, 2007, http://www.fastcompany.com/60374/working-enemy.

39. Knudson, "Environment Inc."

40. Center for Biological Diversity, "Human Population Growth and Extinction," http://www.biologicaldiversity.org/programs/population_and_sustainability /extinction/.

41. Bob Beck, "Wyoming's Congresswoman Wants to Reform the Endangered Species Act," Wyoming Public Radio, August 23, 2013, http://wyomingpub licmedia.org/post/wyoming-s-congresswoman-wants-reform-endangered-spe-cies-act.

42. Savingspecies.org.

43. Brad Plumber, "The World Is on the Brink of a Mass Extinction. Here's How to Avoid That," Vox Media, June 11, 2014, http://www.vox.com/2014/6 /11/5797636/the-world-is-facing-a-major-extinction-crisis-here-are-ways-to -avoid.

44. Toby Wolpe, "HP's Earth Insights Deploys Big Data Tech against Eco Threats," *ZDNet*, December 10, 2013, http://www.zdnet.com/hps-earth-insights-de ploys-big-data-tech-against-eco-threats-7000024118/.

45. HP Press Release, "Revolutionary Big Data Environmental Analysis Indicates Declining Animal Populations in Studied Tropical Forests," December 10, 2013, http://www8.hp.com/us/en/hp-news/press-release.html?id=1536855# .VC28NRbzZuA.

46. Ibid.

47. Robert Costanza, Ralph D'Arge, Rudolph de Groot, et al., "The Value of the World's Ecosystem Services and Natural Capital," *Nature* 387 (May 15, 1997), pp. 253–60, http://www.esd.ornl.gov/benefits_conference/nature_paper.pdf.

48. Walter V. Reid, Harold A. Mooney, Angela Cropper, et al., "Ecosystems and Human Well Being, Millennium Ecosystem Assessment," 2005, http://www .millenniumassessment.org/documents/document.356.aspx.pdf.

49. Gretchen Daily and Katherine Ellison, *The New Economy of Nature: The Quest to Make Conservation Profitable* (Washington, DC: Island Press, 2002), p. 4.

Chapter 4

1. U.S. Geographical Survey, "California's Central Valley Groundwater Study: A Powerful New Tool to Assess Water Resources in California's Central Valley," http://pubs.usgs.gov/fs/2009/3057/.
2. Julia Lurie, "California Farms Are Sucking Up Enough Groundwater to Put Rhode Island 17 Feet Under," *Mother Jones*, July 16, 2014, http://www.mother jones.com/blue-marble/2014/07/california-drought-report-economy-ground water.
3. Richard Howitt, Josué Medellín-Azuara, Duncan MacEwan, et al., "Economic Analysis of the 2014 Drought for California Agriculture," Center for Watershed Sciences, University of California, Davis, July 23, 2014, p. ii, https://watershed.ucdavis.edu/files/biblio/DroughtReport_23July2014_0.pdf.
4. Steve Adler, "Farm Water Shortages Could Linger in 2014," *AgAlert*, June 26, 2013, http://www.agalert.com/story/?id=5702.
5. Robert Rodriguez, "Drought Drying Up Small Central Valley Farmers' Future," *The Fresno Bee*, July 19, 2014, http://www.fresnobee.com/2014/07/19/4032162_drought-drying-up-small-farmers.html?rh=1.
6. Michelle Nijhuis, "Amid Drought, New California Law Will Limit Groundwater for First Time," *National Geographic*, September 17, 2014, http://news.nationalgeographic.com/news/2014/09/140917-california-groundwater-law-drought-central-valley-environment-science/.
7. Brett Walton, "California's Dogged Drought Cutting Off Water Supplies to State's Poor," *Circle of Blue*, August 20, 2014, http://www.circleofblue.org/waternews/2014/world/californias-dogged-drought-cutting-water-supplies-states-poor/.
8. Scott Kraft, "Drinking Water Crisis: A Californian Town Fights Back," *Los Angeles Times*, November 7, 2010, http://latimesblogs.latimes.com/greenspace/2010%20/11/drinking-water-nitrates-california-agricultural-runoff.html.
9. Katie Paul, "Water Shortage Wilts Calif.'s San Joaquin Valley," *Newsweek*, April 24, 2012, http://www.newsweek.com/water-shortage-wilts-califs-san-joaquin-valley-78823.
10. Nijhuis, "Amid Drought, New California Law Will Limit Groundwater for First Time."
11. Lurie, "California Farms Are Sucking Up Enough Groundwater to Put Rhode Island 17 Feet Under."
12. Stephen D. Simpson, "Top Agricultural Producing Countries," *Investopedia*, July 12, 2012, http://www.investopedia.com/financial-edge/0712/top-agricultural-producing-countries.aspx.

13. Glenn Schaible and Marcel Aillery, "Water Conservation in Irrigated Agriculture: Trends and Challenges in the Face of Emerging Demands," U.S. Department of Agriculture, September 2012, http://www.ers.usda.gov/publications /eib-economic-information-bulletin/eib99.aspx#.VCHIOhbzZuA.

14. U.S. Geological Survey, "Total Water Use in the United States, 2005," http:// water.usgs.gov/edu/wateruse-total.html.

15. Food and Agriculture Organization of the United Nations and Earthscan, "The State of the World's Land and Water Resources for Food and Agriculture: Managing Systems at Risk," 2011, http://www.fao.org/docrep/017/i1688e /i1688e.pdf.

16. United Nations, World Water Day 2013: International Year of Water Cooperation, "Facts & Figures," http://www.unwater.org/water-cooperation-2013 /water-cooperation/facts-and-figures/en/.

17. Jonathan A. Foley, "Can We Feed the World and Sustain the Planet?" *Scientific American*, October 12, 2011, http://www.scientificamerican.com/article/can -we-feed-the-world/.

18. Ali Partovi, "Food Is the New Frontier in Green Tech," *TechCrunch*, April 24, 2011, http://techcrunch.com/2011/04/24/ali-partovi-fix-food/.

19. American Experience, "Timeline of Farming in the U.S.," http://www.pbs.org /wgbh/amex/trouble/timeline/.

20. Nancy M. Trautmann, Keith S. Porter, and Robert J. Wagenet, "Modern Agriculture: Its Effects on the Environment," Cornell University Cooperative Extension, http://psep.cce.cornell.edu/facts-slides-self/facts/mod-ag-grw85.aspx.

21. Paul K. Conkin, *A Revolution Down on the Farm: The Transformation of American Agriculture since 1929* (Lexington: University of Kentucky Press, 2008), p. 98.

22. U.S. Department of Commerce, Bureau of the Census, "American Cotton Supply and Distribution, Season of 1916–17," p. 28, http://books.google.com /books?id=0jkqAAAAYAAJ&pg=PA28#v=onepage&q&f=falsem.

23. National Cotton Council of America, "Cotton Production Costs and Returns: United States," http://www.cotton.org/econ/cropinfo/costsreturns/usa.cfm.

24. Trautmann, Porter, and Wagenet, "Modern Agriculture: Its Effects on the Environment."

25. U.S. Geological Survey, "Pesticides in the Nation's Streams and Ground Water, 1992–2001: A Summary," March 2006, http://pubs.usgs.gov/fs/2006/3028/.

26. U.S. Environmental Protection Agency, "Healthy Watersheds," http://water .epa.gov/polwaste/nps/watershed/index.cfm.

27. Conkin, *A Revolution Down on the Farm*, p. 171.

28. Schaible and Aillery, "Water Conservation in Irrigated Agriculture."

29. "Big-Fish Stocks Fall 90 Percent since 1950, Study Says," *National Geographic News,* May 15, 2003, http://news.nationalgeographic.com/news/2003/05/0515 _030515_fishdecline.html.

30. Zeke Grader, "Greener than the Greens: Why the Economic Interests of Fishermen Make Them More Environmentalist than Environmental Groups," *Fishermen's News,* April 1999, the Pacific Coast Federation of Fishermen's Associations, http://www.pcffa.org/fn-apr99.htm.

31. Foley, "Can We Feed the World and Sustain the Planet?"

32. CBC Television, "David Suzuki Speaks Out Against Genetically Modified Food," October 17, 1999, http://www.cbc.ca/player/Digital+Archives /Science+and+Technology/Biotechnology/ID/1936218986/.

33. Charles M. Benbrook, "Impacts of Genetically Modified Engineered Crops on Pesticide Use in the U.S.: The First Sixteen Years," *Environmental Sciences Europe,* September 28, 2012, http://www.enveurope.com/content/24/1/24.

34. Tim Folger, "The Next Green Revolution, *National Geographic Magazine*, October 2014, http://www.nationalgeographic.com/foodfeatures/green-revolu tion/.

35. Allison Aubrey, "Test Your Food IQ: Do We Need More Farms to Grow Fruits and Veggies for All?" National Public Radio, October 17, 2012, http://www .npr.org/blogs/thesalt/2012/10/17/163085279/test-your-food-iq-do-we -need-more-farms-to-grow-fruits-and-veggies-for-all.

36. Marshall J. English, Kenneth J. Solomon, and Glenn J. Hoffman, "A Paradigm Shift in Irrigation Management," *Journal of Irrigation and Drainage Engineering,* September–October 2002, pp. 267–77.

37. University of Massachusetts at Amherst, "An Overview of Drip Irrigation," https://extension.umass.edu/vegetable/articles/overview-drip-irrigation.

38. Lined Canals, "Advantages and Disadvantages of Canal Lining," January 29, 2010, http://linedcanals.wordpress.com/tag/evaporation/.

39. Dickson P. Despommier, "A Farm on Every Floor," *The New York Times,* August 23, 2009, http://www.nytimes.com/2009/08/24/opinion/24Despommier.html?_r=0.

40. U.S. Geological Survey, "Irrigation Techniques," http://water.usgs.gov/edu /irmethods.html.

41. Schaible and Aillery, "Water Conservation in Irrigated Agriculture."

42. Foley, "Can We Feed the World and Sustain the Planet?"

43. John Carey, "Precision Conservation," University of Washington, *Conservation,* September 9, 2013, http://conservationmagazine.org/2013/09/precision-con servation/.

44. Ibid.

45. Lucy Allen, "Smart Irrigation Scheduling: Tom Rogers' Almond Ranch," Pacific Institute Farm Water Success Stories, http://pacinst.org/wp-content/up loads/sites/21/2013/02/smart_irrigation_scheduling3.pdf.

46. Allie Arp, "Utilizing Satellite Imagery in On-Farm Network Trials," On-Farm Network, *Advance*, September 25, 2014, http://www.isafarmnet.com/Advance /AdvanceSeptember25_2014_1.pdf.

47. Cleanwater Iowa, "Tour Highlights Environmental Work of Farmers," November 15, 2003, http://www.cleanwateriowa.org/article.aspx?id=22&Tour+highli ghts+environmental+work+of+farmers+|+Iowa+Soybean+Association.

48. Kathleen Matterson, "Solution in the Soil? Farming for a Cleaner Gulf," *Harvest Public Media*, October 21, 2010, http://harvestpublicmedia.org/article /solution-soil-farming-cleaner-gulf.

49. California Department of Water Resources, "CIMIS Overview," http://ww wcimis.water.ca.gov/Default.aspx#.

50. Julie Menter, "Helping U.S. Farmers Increase Production and Protect the Land," *Environment 360*, July 5, 2012, http://e360.yale.edu/feature/helping _us_farmers_increase_production_and_protect_the_land/2549/.

51. Foley, "Can We Feed the World and Sustain the Planet?"

52. Savory Institute, "Allan Savory," http://www.savoryinstitute.com/about-us /our-team/allan-savory/.

53. Allan Savory, "How to Fight Desertification and Reverse Climate Change," TED Talk, February 2013, https://www.ted.com/talks/allan_savory_how_to _green_the_world_s_deserts_and_reverse_climate_change.

54. United Nations, World Day to Combat Desertification, "Desertification," http://www.un.org/en/events/desertificationday/background.shtml.

55. Savory, "How to Fight Desertification and Reverse Climate Change."

56. "Can Organic Feed the World? New Study Sheds Light on Debate over Organic vs. Conventional Agriculture," *ScienceDaily*, April 25, 2012, http://www .sciencedaily.com/releases/2012/04/120425140114.htm.

57. Statista.com, "Worldwide Agrochemical Market Revenue in 2013 and 2018 (in Billion U.S. Dollars)," http://www.statista.com/statistics/311957/global -agrochemical-market-revenue-projection/.

58. Field to Market, "Boone River Watershed Fieldprint Project," https://www .fieldtomarket.org/fieldprinting-projects/files/Fieldprint_Project_Fact_Sheet _Boone_River.pdf.

Chapter 5

1. James Tapper, "Coke May Have Won the World Cup, but It's Losing in India," *Global Post*, June 23, 2014, http://www.globalpost.com/dispatch/news /regions/asia-pacific/india/140623/coca-cola-coke-business-economy.

2. Avantika Chilkoti, "Water Shortage Shuts Coca-Cola Plant in India," CNBC, June 20, 2014, http://www.cnbc.com/id/101775300#.

3. India Resource Center, "Authorities Cancel License for Coca-Cola's Mehdiganj Plant," June 18, 2014, http://www.indiaresource.org/news/2014/1020.html.

4. Tapper, "Coke May Have Won the World Cup, but It's Losing in India."

5. Intel website, "Our Wastewater Collection Systems," http://www.intel.com /content/www/us/en/environment/environment-wastewater-collection-story .html.

6. Matthew Power, "Peak Water: Aquifers and Rivers Are Running Dry. How Three Regions Are Coping," *Wired Magazine*, April 21, 2008, http://archive.wired .com/science/planetearth/magazine/16-05/ff_peakwater?currentPage=all.

7. PR Newswire, "Soft Drinks: Global Industry Guide, MarketLine Industry Guides," July 2, 2013, http://www.prnewswire.com/news-releases/soft-drinks -global-industry-guide-marketline-industry-guides-213979741.html.

8. WaterFootprint.org, "Product Water Footprint—Soft Drinks," http://www.wa terfootprint.org/?page=files/Softdrinks.

9. "Pure Water, Semiconductors, and the Recession," *Global Water Intelligence* 10, no. 10 (October 2009), http://www.globalwaterintel.com/archive/10/10/mar ket-insight/pure-water-semiconductors-and-the-recession.html.

10. Intel 2009 Proxy Statement, "Policy on Human Right to Water," http://www .intc.com/intelproxy2009/proposal_7/index.html.

11. National Geographic, "Change the Course: Water Footprint Calculator," http://environment.nationalgeographic.com/environment/freshwater/change -the-course/water-footprint-calculator/.

12. Tracey Schelmetic, "Down the Drain: Industrial Water Use," Thomasnet.com, April 10, 2012, http://news.thomasnet.com/IMT/2012/04/10/down-the -drain-industry-water-use.

13. Dennis Nelson, "Six Ways We All Use Water Without Knowing It," *Mother Nature Network*, http://www.mnn.com/food/beverages/sponsorstory/six-ways -we-all-use-water-without-knowing-it.

14. Carbon Disclosure Project, *Collective Responses to Rising Water Challenges*, CDP Global Water Report 2012, https://www.cdp.net/CDPResults/CDP-Water -Disclosure-Global-Report-2012.pdf.

15. U.S. Geological Survey, "Industrial Water Use," http://water.usgs.gov/edu /wuin.html.

16. UNESCO, *Water for People, Water for Life*, United Nations Water Development Report, 2003, http://unesdoc.unesco.org/.

17. The United Nations World Water Development Report 2014, *Facing the Challenges*, Volume 2, p. 70, http://unesdoc.unesco.org/images/0022/002257/225741E .pdf#page=153.

18. Patagonia Website, "The Responsible Economy," http://www.patagonia.com /us/patagonia.go?assetid=1865.

19. Carbon Disclosure Project, *Moving Beyond Business as Usual: A Need for a Step Change in Water Risk Management*, CDP Global Water Report 2013, p. 10, https://www.cdp.net/CDPResults/CDP-Global-Water-Report-2013.pdf.

20. Andrew Winston, "Resilience in a Hotter World," *Harvard Business Review*, April 2014, https://hbr.org/2014/04/resilience-in-a-hotter-world.

21. Jennifer Elks, "Consumers, Activists Declare Victory as General Mills Commits to Non-GMO Cheerios," SustainableBrands.com, January 3, 2014, http://www.sustainablebrands.com/news_and_views/marketing_communica tions/jennifer_elks/consumers_activists_declare_victory_general_mi.

22. Dominic Barton and Mark Wiseman, "Focusing Capital on the Long Term: Big Investors Have an Obligation to End the Plague of Short-Termism," *Harvard Business Review*, January–February 2014, pp. 48–55, http://www .top1000funds.com/wp-content/uploads/2014/01/Focusing-Capital-on-the -Long-Term.pdf.

23. John Elkington and Jochen Zeitz, *The Breakthrough Challenge: 10 Ways to Connect Today's Profits to Tomorrow's Bottom Line* (San Francisco: Jossey-Bass, 2014).

24. Susan Adams, "Can a New Corporate Structure Get Companies to Do Good by Not Always Putting Shareholders First?" Forbes.com, March 25, 2010, http://web.archive.org/web/20100328155935/http://www.forbes.com /forbes/2010/0412/rebuilding-b-lab-corporate-citizenship-green-incorpora tion-mixed-motives.html.

25. James Surowiecki, "Companies with Benefits," *The New Yorker*, August 4, 2014, http://www.newyorker.com/magazine/2014/08/04/companies-benefits.

26. Hannah Clark Steiman, "A New Kind of Company: A 'B' Corporation," Inc. com, July 1, 2007, http://www.inc.com/magazine/20070701/priority-a-new -kind-of-company.html.

27. Amy Westervelt, "A New Corporation for a New Economy," BCorporation.net, http://www.bcorporation.net/sites/all/themes/adaptivetheme/bcorp/pdfs/2009 AP-New-Corporation.pdf.

28. B Corporation, "Legislation," http://www.bcorporation.net/what-are-b-corps/legislation.

29. Surowiecki, "Companies with Benefits."

30. Winston, "Resilience in a Hotter World."

31. Yale University, "Global Metrics for the Environment," http://epi.yale.edu/.

32. Organisation for Economic Co-operation and Development, "Moving towards a Common Approach on Green Growth Indicators," April 2012, http://issuu.com/ggkp/docs/ggkp_moving_towards_a_common_approa.

33. PwC website, "Total Impact Measurement and Management," http://www.pwc.com/gx/en/sustainability/publications/total-impact-measurement-management/index.jhtml.

34. Sarah Sorenson, *The Sustainable Network: The Accidental Answer for a Troubled Planet* (Sebastopol, CA: O'Reilly Media, 2009), p. 147.

35. Pilita Clark, "A World without Water," *Financial Times*, July 14, 2014, http://www.ft.com/cms/s/2/8e42bdc8-0838-11e4-9afc-00144feab7de.html#slide0.

36. Trucost.com, "Case Study: Rethinking Profit and Loss, Yorkshire Water," 2014, http://www.trucost.com/_uploads/publishedResearch/Yorkshire%20Water%20case%20study_2014.pdf.

37. United Nations, *The CEO Water Mandate: An Initiative by Business Leaders in Partnership with the International Community*, http://ceowatermandate.org/files/Ceo_water_mandate.pdf.

38. Joel Makower, "Color It Green: Nike to Adopt Waterless Textile Dyeing," *GreenBiz*, February 7, 2012, http://www.greenbiz.com/blog/2012/02/07/color-it-green-nike-adopt-waterless-textile-dyeing.

39. Lynn S. Paine, "Sustainability in the Board Room," *Harvard Business Review*, July 2014, p. 93, https://hbr.org/2014/07/sustainability-in-the-boardroom.

40. Charles Fishman, "Why GE, Coca-Cola and IBM Are Getting into the Water Business," *Fast Company*, April 11, 2011, http://www.fastcompany.com/1739772/why-ge-coca-cola-and-ibm-are-getting-water-business.

41. Carbon Disclosure Project, *Moving Beyond Business as Usual*, p. 15.

42. Intel website, "Our Wastewater Collection Systems."

43. Ford Motor Company website, "Sustainability Report 2013/2014: Water," http://corporate.ford.com/microsites/sustainability-report-2013-14/water.html.

44. The Water Council, "MillerCoors Receives 2013 U.S. Water Prize," February 28, 2013, http://thewatercouncil.wordpress.com/2013/02/28/millercoors-receives-water-prize/.

45. Unilever Progress Report 2012, "Unilever Sustainable Living Plan," pp. 32–

22, http://www.unilever.com/images/USLP-Progress-Report-2012-FI_tcm13
-352007.pdf.

46. Adele Peters, "What's Behind the Big Sustainability Push at Unilever?" *Green-Biz*, June 29, 2011, http://www.greenbiz.com/blog/2011/06/29/whats-behind
-big-sustainability-push-unilever?page=0%2C0.

47. Peter Kelly-Detwiler, "Water FX Sees Solar Desalination as One Way to Address the World's Water Problem," *Forbes*, January, 7, 2014, http://www.forbes
.com/sites/peterdetwiler/2014/01/07/waterfx-sees-solar-desalination-as-one-way-to-address-the-worlds-water-problem/.

48. Coca-Cola Water Stewardship & Replenishment Report, "Collaborating to Replenish the Water We Use," http://www.coca-colacompany.com/collaborat
ing-to-replenish-the-water-we-use.

49. "Coca-Cola 2013/2014 Sustainability Report," p. 48, http://assets.coca-cola
company.com/0a/b5/ece07f0142ce9ccc4504e28f1805/2013-2014-coca-cola
-sustainability-report-pdf.pdf.

50. Sarah Wade, "Agua Dulce: Renewing Freshwater Resources and Strengthening Livelihoods around the Mesoamerican Reef," *World Wildlife Magazine*, Spring 2014, http://www.worldwildlife.org/magazine/issues/spring-2014/articles/agua-dulce.

51. Burt Helm, "Climate Change's Bottom Line," *The New York Times*, January 31, 2015, http://www.nytimes.com/2015/02/01/business/energy-environment/cli
mate-changes-bottom-line.html?_r=1.

52. Beverage Industry Environmental Roundtable website, http://www.bieround
table.com/.

53. Beverage Industry Environmental Roundtable, *Managing Water-Related Business Risks & Opportunities in the Beverage Sector*, November 2012, http://media
.wix.com/ugd/49d7a0_f49252ae57154a7baefbd0c314e311f1.pdf.

54. Nestlé, "Creating Shared Value Full Report 2013," p. 172, http://storage.nes
tle.com/Interactive_CSV_Full_2013/index.html#189/z.

55. Clark, "A World without Water."

56. Carbon Disclosure Project, *Moving Beyond Business as Usual*, p. 15.

57. Ceres, "Gaining Ground: Corporate Progress on the Ceres Roadmap for Sustainability," 2014, p. 6, http://www.ceres.org/resources/reports/gaining-ground
-corporate-progress-on-the-ceres-roadmap-for-sustainability/view.

Chapter 6

1. Richard Waters, "An Exclusive Interview with Bill Gates," *Financial Times Magazine*, November 1, 2013, http://www.ft.com/intl/cms/s/2/dacd1f84-41bf
-11e3-b064-00144feabdc0.html#axzz3IKIcRv00.

2. Liz Stinson, "This Revolutionary Cooler Could Save Millions of Lives," *Wired*, June 18, 2013, http://www.wired.com/2013/06/how-to-design-a-life-saving-device/.

3. Bill & Melinda Gates Foundation website, "Reinvent the Toilet Challenge," http://www.gatesfoundation.org/What-We-Do/Global-Development/Reinvent-the-Toilet-Challenge.

4. Waters, "An Exclusive Interview with Bill Gates."

5. Ibid.

6. World Health Organization, "Factsheet on the World Malaria Report 2013," December 2013, http://www.who.int/malaria/media/world_malaria_report_2013/en/.

7. *CBS News*, "Amid Ebola Outbreak, Bill Gates Pledges More Money to Fight Malaria," November 2, 2014, http://www.cbsnews.com/news/amid-ebola-outbreak-gates-foundation-boosts-aid-to-stamp-out-malaria/.

8. Charles Strum, "Belleville Journal; Restoring Heritage and Raising Hopes for Future," *The New York Times*, March 2, 1992, http://www.nytimes.com/1992/03/02/nyregion/belleville-journal-restoring-heritage-and-raising-hopes-for-future.html?src=pm.

9. The Rockefeller Foundation website, "Eradicating Hookworm," http://rockefeller100.org/exhibits/show/health/eradicating-hookworm.

10. Bradley Lubben and James Pease, "Conservation and the Agricultural Act of 2014," *Choices*, 2nd quarter 2014, http://www.choicesmagazine.org/choices-magazine/theme-articles/deciphering-key-provisions-of-the-agricultural-act-of-2014/conservation-and-the-agricultural-act-of-2014.

11. USDA Natural Resources Conservation Service website, "Environmental Quality Incentives Program," http://www.nrcs.usda.gov/wps/portal/nrcs/main/national/programs/financial/eqip/.

12. Environmental Working Group, *Untapped: How Farm Bill Conservation Programs Can Do More to Clean Up California's Water*, October 2013, http://static.ewg.org/pdf/2013_California_EQIP_Report.pdf?_ga=1.38501482.853323352.1415651832.

13. Michael E. Porter and Mark R. Kramer, "Philanthropy's New Agenda: Creating Value," *Harvard Business Review*, November–December 1999, p. 122, http://static.ewg.org/pdf/2013_California_EQIP_Report.pdf?_ga=1.38501482.853323352.1415651832.

14. Steven H. Goldberg, *Billions of Drops in Millions of Buckets: Why Philanthropy Doesn't Advance Social Progress* (Hoboken, NJ: John Wiley & Sons, 2009), p. 7.

15. Christine W. Letts, William P. Ryan, and Allen S. Grossman, "Virtuous Capi-

tal: What Foundations Can Learn from Venture Capitalists," *Harvard Business Review*, March 1997, p. 5, https://hbr.org/1997/03/virtuous-capital-what -foundations-can-learn-from-venture-capitalists.

16. Matthew Bishop and Michael Green, *Philanthrocapitalism: How the Rich Can Save the World* (New York: Bloomsbury Press, 2008), p. 158.

17. Ibid.

18. Letts, Ryan, and Grossman, "Virtuous Capital," p. 5.

19. Porter and Kramer, "Philanthropy's New Agenda," p. 129.

20. Ruth McCambridge, "Small Survey of Foundations on 'Hot Button Issues' Shows Little Hysteria," *Nonprofit Quarterly*, November 8, 2013, https://non profitquarterly.org/philanthropy/23212-small-survey-of-foundations-on-hot -button-issues-shows-little-hysteria.html.

21. Goldberg, *Billions of Drops in Millions of Buckets*, p. 15.

22. Letts, Ryan, and Grossman, "Virtuous Capital," p. 6.

23. Porter and Kramer, "Philanthropy's New Agenda," p. 122.

24. Paul Brest and Kelly Born, "When Can Impact Investing Create Real Impact?" *Stanford Social Innovation Review*, Fall 2013, http://www.ssireview.org/up_for _debate/article/impact_investing.

25. National Philanthropic Trust website, "Charitable Giving Statistics," http:// www.nptrust.org/philanthropic-resources/charitable-giving-statistics/.

26. Alex Daniels, "Donors Care More about How Money Is Spent than Results," *The Chronicle of Philanthropy*, October 28, 2014, http://philanthropy.com/ar ticle/Donors-Care-More-About-How/149669?cid=megamenu.

27. "The Business of Giving," *The Economist*, February 23, 2006, http://www .economist.com/node/5517605.

28. Andy Serwer, "The Legend of Robin Hood," *Fortune Magazine*, September 8, 2006, http://archive.fortune.com/magazines/fortune/fortune_archive/2006 /09/18/8386204/index.htm.

29. Robin Hood website, "Targeting Poverty in New York City," https://www.rob inhood.org/impact#section-2.

30. CNBC, "Robin Hood Donors Raise $60 Million at Annual Gala," May 13, 2014, http://www.cnbc.com/id/101669752#.

31. Serwer, "The Legend of Robin Hood."

32. Matt Bannick and Eic Hallstein, "Learning from Silicon Valley: How the Omidyar Network Uses a Venture Capital Model to Measure and Evaluate Effectiveness," *Stanford Social Innovation Review*, Summer 2012, http://www .ssireview.org/articles/entry/learning_from_silicon_valley1.

33. Omidyar Network website, "Investees," https://www.omidyar.com/investees.

34. Peg Tyre, "Beyond School Supplies: How DonorsChoose Is Crowdsourcing Real Education Reform," *Fast Company*, March 2014, http://www.fastcom pany.com/3025597/donorschoose-hot-for-teachers.

35. DonorsChoose website, "How It Works," http://www.donorschoose.org/about.

36. Dan Gilgoff, "Judith Rodin: Rockefeller Foundation Head Changes the Charity and the World," *US News & World Report*, October 22, 2009, http://www .usnews.com/news/best-leaders/articles/2009/10/22/judith-rodin-rockefeller -foundation-head-changes-the-charity-and-the-world.

37. Judith Rodin, "Innovations in Finance for Social Impact," The Rockefeller Foundation Blog, September 5, 2014, http://www.rockefellerfoundation.org /blog/innovations-finance-social-impact.

38. Wharton University of Pennsylvania website, "Impact Investing: Judith Rodin Takes On the Naysayers," June 9, 2014, http://knowledge.wharton.upenn .edu/article/putting-markets-work-profit-global-good/.

39. Judith Rodin, "Innovation for the Next 100 Years," *Stanford Social Innovation Review*, Summer 2013, http://www.ssireview.org/articles/entry/innovation _for_the_next_100_years.

40. New York City Acquisition Fund website, http://www.nycacquisitionfund .com/.

41. Phil Bolton, "Social Entrepreneur Looks to Light Up the World," *Global Atlanta*, August 20, 2009, http://www.globalatlanta.com/article/17522/social -entrepreneur-looks-to-light-up-the-world/.

42. d.light website, http://www.dlight.com/about-us/.

43. David W. Chen, "Goldman to Invest in City Jail Program, Profiting if Recidivism Falls Sharply," *The New York Times*, August 2, 2012, http://www.nytimes .com/2012/08/02/nyregion/goldman-to-invest-in-new-york-city-jail-pro gram.html?pagewanted=all&_r=0.

44. Jon Hartley, "Social Impact Bonds Are Going Mainstream," *Forbes*, September 15, 2014, http://www.forbes.com/sites/jonhartley/2014/09/15/social-impact -bonds-are-going-mainstream/.

45. Chen, "Goldman to Invest in City Jail Program."

46. Paul Sullivan, "Investing to Make a Difference Is Gaining Ground," *The New York Times*, September 5, 2014, http://www.nytimes.com/2014/09/06/your -money/asset-allocation/investing-to-make-a-difference-is-gaining-ground -.html?emc=eta1&_r=0.

47. HIP Investor website, http://hipinvestor.com/.

48. Sullivan, "Investing to Make a Difference Is Gaining Ground."

49. Chris Witkowsky, "Agribusiness Fund Seeks $250 Mln for Neglected Market;

Washington, Maine Pensions on Board," *PE HUB*, September 3, 2014, https://www.pehub.com/2014/09/agri-business-fund-seeks-250-mln-for-neglected-market-washington-maine-pensions-on-board/.

50. Equilibrium Capital, "The Investment Opportunity in Permanent Crops," http://vimeo.com/69972757.

51. J. Fullerton, "Regenerative Capitalism: How Universal Principles and Patterns Will Shape Our New Economy," Capital Institute, 2015, http://capital institute.org/wp-content/uploads/2015/04/2015-Regenerative-Capitalism-4-20-15-final.pdf.

52. Willamette Partnership, "Joint Statement for an Ecosystem Credit Accounting System," September 2009, http://willamettepartnership.org/wp-content/uploads/2014/10/Counting-on-the-Environment-Joint-Statement_2009.pdf.

53. Willamette Partnership, "Regional Recommendations on Water Quality Trading," http://willamettepartnership.org/success-stories/regional-recommendations-water-quality-trading/.

Chapter 7

1. Nick Allen, "The Race to Stop Las Vegas from Running Dry," *The Telegraph*, June 28, 2014, http://www.telegraph.co.uk/news/worldnews/northamerica/usa/10932785/The-race-to-stop-Las-Vegas-from-running-dry.html.

2. "Water Worries: The Drying of the West," *The Economist*, January 27, 2011, http://www.economist.com/node/18013810.

3. John M. Glionna, "Drought—and Neighbors—Press Las Vegas to Conserve Water," *Los Angeles Times*, April 20, 2014, http://www.latimes.com/nation/la-na-las-vegas-drought-20140421-story.html#page=1.

4. Ibid.

5. Frances Weaver, "The Unprecedented Water Crisis of the American Southwest," *The Week*, February 1, 2014, http://theweek.com/articles/451876/un precedented-water-crisis-american-southwest.

6. Allen, "The Race to Stop Las Vegas from Running Dry."

7. Weaver, "The Unprecedented Water Crisis of the American Southwest."

8. Michael Wines, "Colorado River Drought Forces a Painful Reckoning for States," *The New York Times*, January 5, 2014, http://www.nytimes.com/2014/01/06/us/colorado-river-drought-forces-a-painful-reckoning-for-states.html?_r=0; Sarah Zielinski, "The Colorado River Runs Dry," *Smithsonian Magazine*, October 2010, http://www.smithsonianmag.com/science-nature/the-colorado-river-runs-dry-61427169/?no-ist.

9. Glionna, "Drought—and Neighbors—Press Las Vegas to Conserve Water."

10. Weaver, "The Unprecedented Water Crisis of the American Southwest."

11. Glionna, "Drought—and Neighbors—Press Las Vegas to Conserve Water."

12. Keith Schneider, "Australia's Food Bowl, Like the World's, Is Drying Up," *Circle of Blue*, March 9, 2009, http://www.circleofblue.org/waternews/2009 /world/australia-drought-water-warning/.

13. Murray–Darling Basin Authority, "Guide to the Proposed Basin Plan: Volume 1, Overview," September 2, 2011, p. XV, http://www.mdba.gov.au/sites/de fault/files/archived/Guide_to_the_Basin_Plan_Volume_1_web.pdf.

14. Schneider, "Australia's Food Bowl, Like the World's, Is Drying Up."

15. Nick Bryant, "'Big Dry' Turns Farms into Desert," *BBC News, Australia*, August 31, 2008, http://news.bbc.co.uk/2/hi/asia-pacific/7577528.stm.

16. Schneider, "Australia's Food Bowl, Like the World's, Is Drying Up."

17. Murray–Darling Basin Authority, "Guide to the Proposed Basin Plan: Volume 1, Overview," August 10, 2010, p. 13, http://www.mdba.gov.au/kid/guide/in dex.php.

18. Murray–Darling Basin Ministerial Council, "Native Fish Strategy for the Murray–Darling Basin 2003–2013," May 2003, p. 1, http://www.mdba.gov.au /sites/default/files/Fish-Strat_ful_2003-13.pdf; Fran Sheldon, "Australia Begins Large-Scale Plan to Rehydrate Declining River System," *The Guardian*, January 14, 2013, http://www.theguardian.com/sustainable-business/austra lia-murray-darling-basin-plan-river-agriculture.

19. Schneider, "Australia's Food Bowl, Like the World's, Is Drying Up."

20. Keith Bradsher, "A Drought in Australia, A Global Shortage of Rice," *New York Times*, April 17, 2008, http://www.nytimes.com/2008/04/17/business /worldbusiness/17warm.html?pagewanted=all&_r=0.

21. Mark Turrell, "Australia's Epic Drought, a Warning of Global Water Scarcity, 'The Biggest Dry,'" Imaginatik, March 10, 2009, http://imaginatik.com/news /australias-epic-drought-warning-global-water-scarcity-biggest-dry.

22. Robert Draper, "Australia's Dry Run," *National Geographic*, April 2009, http:// ngm.nationalgeographic.com/2009/04/murray-darling/draper-text.

23. S.A. Weather and Disaster Information Service, "Drought a Disaster for Adelaide Residents," January 1, 2010, http://saweatherobserver.blogspot.com /2010/01/drought-disaster-for-adelaide-residents.html.

24. Draper, "Australia's Dry Run."

25. "South Australia Drought Worsens," *BBC News*, July 10, 2008, http://news.bbc .co.uk/nol/ukfs_news/hi/newsid_7490000/newsid_7499000/7499036.stm.

26. Murray–Darling Basin Authority, "Guide to the Proposed Basin Plan: Volume 1, Overview," p. 4.

27. Australian Government Department of the Environment, "National Water Initiative," http://www.environment.gov.au/topics/water/australian-government-water-leadership/national-water-initiative.

28. Paul Kildea and George Williams, "Journals Excerpt: The Water Act and the Murray–Darling Basin Plan," Thomson Reuters, May 19, 2011, http://sites.thomsonreuters.com.au/journals/2011/05/19/journals-excerpt-the-water-act-and-the-murray-darling-basin-plan/.

29. Murray–Darling Basin Authority, "Development of the Basin Plan," http://www.mdba.gov.au/what-we-do/basin-plan/development.

30. Patrick J. Byrne, "Angry Farmers Burn Murray–Darling Plan," *News Weekly*, October 30, 2010, http://www.newsweekly.com.au/article.php?id=4564.

31. Murray–Darling Basin Authority, "Fact Sheet: The Basin Plan Concept Statement," p. 1, http://www.mdba.gov.au/media-pubs/publications/bp-concept-statement.

32. Murray–Darling Basin Authority, "Guide to the Proposed Basin Plan," p. 43.

33. Oliver Milman, "Plan to Safeguard 'Australia's Food Bowl' Condemned from Both Sides," *The Guardian*, November 28, 2011, http://www.theguardian.com/environment/2011/nov/28/plan-australia-food-bowl.

34. Murray–Darling Basin Authority, "Guide to the Proposed Basin Plan," p. 33.

35. Ibid., pp. xxxi–xxxiii.

36. Australian Government Department of the Environment, "Sustainable Rural Water Use and Infrastructure Program," June 2013, http://laptop.deh.gov.au/water/publications/policy-programs/srwui-factsheet.html.

37. Senator the Hon. Simon Birmingham, Media Release, "$158 Million for On-Farm Irrigation Upgrades," April 1, 2014, http://www.environment.gov.au/minister/birmingham/2014/mr20140401.html.

38. Australian Government, Department of Sustainability, "Environment, Water, Population and Communities, Annual Report 2012–2013," p. 77, http://www.environment.gov.au/system/files/resources/63db8a54-bfcb-429e-93b4-e5efe21a356e/files/dsewpac-annual-report-12-13new.pdf.

39. Australian Government Department of the Environment, "Sustainable Rural Water Use and Infrastructure Program."

40. Murray–Darling Basin Authority, "Guide to the Proposed Basin Plan," p. 38.

41. Yongping Wei, John Langford, Ian R. Willett, et al., "Is Irrigated Agriculture in the Murray Darling Basin Well Prepared to Deal with Reductions in Water Availability?" *Global Environmental Change* 21, no. 3 (August 2011), pp. 906–16.

42. Australian Government National Water Commission, "The Impacts of Water Trading in the Southern Murray–Darling Basin: An Economic, Social and En-

vironmental Assessment," June 2010, http://archive.nwc.gov.au/__data/assets/pdf_file/0019/10783/681-NWC_ImpactsofTrade_web.pdf.

43. David Crowe, "Water Flows Back to Farmers after Coalition Reverses Murray–Darling BuyBacks," *The Australian*, January 20, 2014, http://www.theaustralian.com.au/national-affairs/water-flows-back-to-farmers-after-coalition-reverses-murraydarling-buybacks/story-fn59niix-1226805462241.

44. Senator the Hon. Simon Birmingham, Media Release, "$158 Million for On-Farm Irrigation Upgrades."

45. "Murray–Darling Environmental Water Holders Report 2013," p. 5, http://www.environment.gov.au/system/files/resources/187e1491-d72c-41ab-9e82-9528915e1c96/files/mdb-environmental-water-holders-report-2013.pdf.

46. Australian Government, Department of Sustainability, Environment, Water, Population and Communities, "Annual Report 2012–2013," p. 7, http://www.environment.gov.au/system/files/resources/63db8a54-bfcb-429e-93b4-e5efe21a356e/files/dsewpac-annual-report-12-13new.pdf.

47. Wei, Langford, Willett, et al., "Is Irrigated Agriculture in the Murray Darling Basin Well Prepared to Deal with Reductions in Water Availability?"

48. Milman, "Plan to Safeguard 'Australia's Food Bowl' Condemned from Both Sides."

49. Ibid.

50. "Scientists Want 'Manipulated' Basin Plan Scrapped," *ABC News*, January 19, 2012, http://www.abc.net.au/news/2012-01-18/scientists-want-manipulated-basin-plan-scrapped/3781476.

51. Murray–Darling Basin Authority, "Guide to the Proposed Basin Plan," p. xix.

52. Wentworth Group of Concerned Scientists, "National Water Reform," http://wentworthgroup.org/programs/national-water-reform/.

53. "Colorado River Law and Policy: Frequently Asked Questions," WaterPolicy.info, p. 19, http://www.waterpolicy.info/projects/CRGI/materials/Colorado%20River%20FAQ%20v1.pdf.

54. U.S. Department of Interior, *Colorado River Interim Guidelines for Lower Basin Shortages and the Coordinated Operations for Lake Powell and Lake Mead*, December 2007, http://www.usbr.gov/lc/region/programs/strategies/RecordofDecision.pdf.

55. International Boundary and Water Commission, "Minute No. 319: Interim International Cooperative Measures in the Colorado River Basin through 2017 and Extension of Minute 318 Cooperative Measures to Address the Continued Effects of the April 2010 Earthquake in the Mexicali Valley, Baja California," November 20, 2012, http://www.ibwc.gov/Files/Minutes/Minute_319.pdf.

Chapter 8

1. Megan Scudellari, "Coastal Command," *The Scientist*, September 1, 2013, http://www.the-scientist.com/?articles.view/articleNo/37163/title/Coastal -Command/.
2. Paul Greenberg, "A River Runs through It," *The American Prospect*, May 22, 2013, http://prospect.org/article/river-runs-through-it.
3. Carol Kaesuk Yoon, "A 'Dead Zone' Grows in the Mississippi," *The New York Times*, January 20, 1998, http://www.nytimes.com/1998/01/20/science/a -dead-zone-grows-in-the-gulf-of-mexico.html.
4. "Blooming Horrible: Nutrient Pollution Is a Growing Problem All along the Mississippi," *The Economist*, June 23, 2012, http://www.economist.com /node/21557365.
5. Nancy Rabalais, "We All Live Downstream (and Upstream)," *Journal of Soil and Water Conservation*, 58, No. 3 (May/June 2003), http://www.jswconline .org/content/58/3/52A.extract.
6. The Nature Conservancy, "America's Great Watershed Initiative," http://www .nature.org/ourinitiatives/habitats/riverslakes/programs/great-rivers-partner ship/americas-great-watershed-initiative-fact-sheet.pdf.
7. Hamline University Center for Global Environmental Education, "Mississippi River Facts & Links," http://cgee.hamline.edu/rivers/resources/river_days/info .html.
8. National Park Service, "Mississippi River Facts," http://www.nps.gov/miss/riv erfacts.htm.
9. Brian Clark Howard, "Mississippi Basin Water Quality Declining despite Conservation," *National Geographic*, April 11, 2014, http://news.national geographic.com/news/2014/04/140411-water-quality-nutrients-pesticides -dead-zones-science/.
10. U.S. Environmental Protection Agency, "Nitrogen and Phosphorus Pollution in the Mississippi River Basin: Findings of the Wadeable Streams Assessment," http://water.epa.gov/type/rsl/monitoring/upload/EPA-MARB-Fact-Sheet -112911_508.pdf.
11. Howard, "Mississippi Basin Water Quality Declining despite Conservation."
12. Brooke Barton and Sarah Elizabeth Clarke, *Water & Climate Risks Facing U.S. Corn Production: How Companies & Investors Can Cultivate Sustainability*, Ceres, June 2014, pp. 45–47, https://www.ceres.org/resources/reports/water -and-climate-risks-facing-u.s.-corn-production-how-companies-and-inves tors-can-cultivate-sustainability/view.

13. "The Mississippi River and the Making of a Dead Zone," Prospect.org, http:// prospect.org/sites/default/files/mississippi-river-web-version.gif.

14. Karl Mathiesen, "Do Dams Destroy Rivers?" *The Guardian*, August 27, 2014, http://www.theguardian.com/environment/live/2014/aug/27/do-dams-de stroy-rivers.

15. Eileen Fretz Shader, "Flood Risk Rising on the Mississippi River?" *American Rivers*, February 14, 2014, http://www.americanrivers.org/blog/flood-risk-ris ing-mississippi-river/#sthash.HqfdzbYG.dpuf.

16. The Nature Conservancy, "The Mississippi River and Its Floodplain," http:// www.nature.org/ourinitiatives/habitats/riverslakes/explore/mississippi-river -and-its-floodplain-restoring-connections-for-people-and.xml.

17. William J. Mitsch, John W. Day Jr., J. Wendell Gilliam, et al., "Reducing Nitrogen Loading to the Gulf of Mexico from the Mississippi River Basin: Strategies to Counter a Persistent Ecological Problem," *Bioscience* 51, no. 5 (2001), pp. 373–388, http://bioscience.oxfordjournals.org/content/51/5/373 .full.pdf+htm.

18. Barton and Clarke, *Water & Climate Risks Facing U.S. Corn Production*, p. 32.

19. Sally Deneen, "Raiding the Bread Basket: Use and Abuse of the Mississippi River Basin," *National Geographic*, January 23, 2012, http://news.nationalgeo graphic.com/news/2012/01/120123-mississippi-river-basin/.

20. The University of British Columbia, "U.S. Rush to Produce Corn-Based Etha- nol Will Worsen 'Dead Zone' in Gulf of Mexico: UBC Study," March 10, 2008, http://news.ubc.ca/2008/03/10/archive-media-releases-2008-mr-08-025/.

21. Patrick J. Kiger, "North Dakota's Salty Fracked Wells Drink More Water to Keep Oil Flowing," *National Geographic*, November 11, 2013, http://news .nationalgeographic.com/news/energy/2013/11/131111-north-dakota-wells -maintenance-water/.

22. Dan Bacher, "Water for Fracking: How Much Does the Oil/Gas Industry Use?" *Daily Kos*, March 23, 2013, http://www.dailykos.com/story/2013/03 /23/1196415/-Water-for-fracking-8-acre-feet-6-721-acre-feet-or-much -much-more.

23. Bill Chameides, "Fracking Water: It's Just So Hard to Clean," *National Geograph- ic*, October 4, 2013, http://energyblog.nationalgeographic.com/2013/10/04 /fracking-water-its-just-so-hard-to-clean/.

24. Marcus Griswold, "Fracking in the Bakken Threatens Missouri River Water- shed Health," National Resources Defense Council Staff Blog, April 9, 2014, http://switchboard.nrdc.org/blogs/mgriswold/fracking_in_the_bakken_con tinu.html.

25. U.S. General Accounting Office, *Oil and Gas Transportation: Department of Transportation Is Taking Actions to Address Rail Safety, but Additional Actions Needed to Improve Pipeline Safety*, August 2014, p. 16, http://www.gao.gov/assets/670/665404.pdf.

26. Alan Bjerga, "Drought-Parched Mississippi River Is Halting Barges," Bloomberg, November 27, 2012, http://www.bloomberg.com/news/2012-11-27/drought-parched-mississippi-river-is-halting-barges.html.

27. Adrian Sainz, "Mississippi Flood of 2011 Caused $2.8B of Economic Damage: Army Corps," *Insurance Journal*, February 27, 2013, http://www.insurance journal.com/news/national/2013/02/27/282875.htm.

28. American Rivers, "America Runs on the Mississippi River," http://www.ameri canrivers.org/rivers/fun/america-runs-on-the-mississippi-river/.

29. Campbell Robertson, "Gulf of Mexico Has Long Been Dumping Site," *The New York Times*, July 29, 2010, http://www.nytimes.com/2010/07/30/us/30gulf.html?pagewanted=all&_r=0.

30. Darryl Fears, "Drought Threatens to Halt Critical Barge Traffic on Mississippi," *The Washington Post*, January 6, 2013, http://www.washingtonpost.com/na tional/health-science/drought-threatens-to-halt-critical-barge-traffic-on-mis sissippidrought-threatens-to-halt-critical-barge-traffic-on-mississippidrought -threatens-to-halt-critical-barge-traffic-on-mississippi/2013/01/06/92498b88 -5694-11e2-bf3e-76c0a789346f_story.html.

31. Craig Cox, Brett Lorenzen, and Soren Rundquist, "Washout: Spring Storms Batter Poorly Protected Soil and Streams," Environmental Working Group, July 3, 2013, http://www.ewg.org/research/spring-storm-batter-midwest-soil -and-streams.

32. U.S. Department of Agriculture National Resources Conservation Service, "Mississippi River Basin Healthy Watershed Initiative," http://www.nrcs.usda .gov/wps/portal/nrcs/detailfull/national/home/?cid=stelprdb1048200.

33. U.S. Department of Agriculture, *Agricultural Resources and Environmental Indicators*, Chapter 6.5, p. 1, http://www.ers.usda.gov/media/873717/wetlands.pdf.

34. Mitsch et al., "Reducing Nitrogen Loading to the Gulf of Mexico from the Mississippi River Basin."

35. U.S. General Accounting Office, *Clean Water Act: Changes Needed if Key EPA Program Is to Help Fulfill the Nation's Water Quality Goals*, December 5, 2013, http://www.gao.gov/products/GAO-14-80.

36. Mississippi River/Gulf of Mexico Watershed Nutrient Task Force, *Gulf Hypoxia Action Plan 2008*, p. 22, http://water.epa.gov/type/watersheds/named /msbasin/upload/2008_8_28_msbasin_ghap2008_update082608.pdf.

37. U.S. Environmental Protection Agency, *Nutrient Pollution: EPA Needs to Work with States to Develop Strategies for Monitoring the Impact of State Activities on the Gulf of Mexico Hypoxic Zone,* September 3, 2014, p. 6, http://www.epa.gov /oig/reports/2014/20140902-14-P-0348.pdf.

38. Mississippi River Collaborative, "EPA Lawsuit," http://www.msrivercollab.org /focus-areas/epa-lawsuit/.

39. Barton and Clarke, *Water & Climate Risks Facing U.S. Corn Production,* p. 49.

40. Mitsch et al., "Reducing Nitrogen Loading to the Gulf of Mexico from the Mississippi River Basin."

41. Ibid.

42. Esri, "What Is GIS?" http://www.esri.com/what-is-gis.

43. The Nature Conservancy, "The Flood's Lingering Effect: New Study Shows Gulf 'Dead Zone' One of the Largest on Record," http://www.nature.org /ourinitiatives/regions/northamerica/areas/gulfofmexico/explore/gulf-of-mex ico-dead-zone.xml.

44. Barton and Clarke, *Water & Climate Risks Facing U.S. Corn Production,* p. 49.

45. Ibid., p. 50.

46. Michelle Perez, Sara Walker, and Cy Jones, *Nutrient Trading in the MRB: A Feasibility Study for Using Large-Scale Interstate Nutrient Trading in the Mississippi River Basin to Help Address Hypoxia in the Gulf of Mexico,* World Resources Institute, http://pdf.wri.org/nutrient_trading_in_mrb_feasibility_study.pdf.

47. Ibid.

48. Evan Branosky, Cy Jones, and Mindy Selman, "Comparison Table of State Nutrient Trading Programs in the Chesapeake Bay Watershed," World Resources Institute, May 2011, p. 6, http://www.wri.org/sites/default/files/comparison _tables_of_state_chesapeake_bay_nutrient_trading_programs.pdf.

49. Perez et al., *Nutrient Trading in the MRB,* p. 11.

50. Barton and Clarke, *Water & Climate Risks Facing U.S. Corn Production,* p. 9.

51. 1 Mississippi, "Dead Zone Size of Connecticut Demands Federal Action," http://1mississippi.org/dead-zone-size-of-connecticut-demands-federal-ac tion/.

52. Mitsch et al., "Reducing Nitrogen Loading to the Gulf of Mexico from the Mississippi River Basin."

Chapter 9

1. Anneke Campbell, "Healing Our City: A Conversation with TreePeople's Andy Lipkis," *LaYoga,* April 2012, pp. 34–37, http://www.layoga-digital.com /layoga/201204#pg38.

2. Jacques Leslie, "Los Angeles, City of Water," *The New York Times*, December 6, 2014, http://www.nytimes.com/2014/12/07/opinion/sunday/los-angeles-city -of-water.html?_r=0.

3. Jared Sichel, "Sukkot, Rain and Andy Lipkis' Vision for L.A.'s Salvation," *Jewish Journal*, October 7, 2014, http://www.jewishjournal.com/cover_story/article /sukkot_rain_and_andy_lipkis_vision_for_l.a.s_salvation_from_the_drought.

4. Leslie, "Los Angeles, City of Water."

5. Willis Harman, *Global Mind Change: The Promise of the 21st Century* (San Francisco: Berrett-Koehler Publishers, 1998), p. viii.

6. Scott London, "Understanding Change: How It Happens, and How to Make It Happen," http://www.scottlondon.com/reports/change.html.

7. Andrew S. Grove, *Only the Paranoid Survive: How to Exploit the Crisis Points That Challenge Every Company* (New York: Doubleday, 1996).

8. Walter Isaacson, *The Innovators: How a Group of Hackers, Geniuses, and Geeks Created the Digital Revolution* (New York: Simon & Schuster, 2014), p. 480.

9. Thom Hartmann, "Transcript: Thom on ADHD, Drugs, Depression, etc.," *Thom Hartmann Program*, March 22, 2006, http://www.thomhartmann.com /blog/2006/03/transcript-thom-adhd-drugs-depression-etc-mar-22-2006.

10. Michael Hammer and James Champy, *Reengineering the Corporation: A Manifesto for Business Revolution* (New York: Harper Collins Publishers, 1993).

11. Peter M. Senge, *The Fifth Discipline: The Art & Practice of the Learning Organization* (New York: Doubleday, 1990).

12. Emily Esfahani Smith and Jennifer L. Aaker, "Millennial Searchers," *The New York Times*, November 30, 2013, http://www.nytimes.com/2013/12/01/opin ion/sunday/millennial-searchers.html?pagewanted=all&_r=0.

13. AMP Agency and Cone, Inc., "The Millennial Generation: Pro-Social and Empowered to Change the World," *GreenBook*, April 23, 2010, http://www .greenbook.org/marketing-research/millennial-cause-study.

14. Oakley Website, "Oakley History," http://in.oakley.com/innovation/history.

Further Reading

Adams, John H., and Patricia Adams. *A Force for Nature: The Story of NRDC and Its Fight to Save Our Planet*. San Francisco: Chronicle Books, 2010.

Arrillaga-Andreessen, Laura. *Giving 2.0: Transform Your Giving and Our World*. San Francisco: Jossey-Bass, 2012.

Barnett, Cynthia. *Blue Revolution: Unmaking America's Water Crisis*. Boston: Beacon Press, 2011.

Berry, Wendell. *Another Turn of the Crank*. Berkeley, CA: Counterpoint Press, 1996.

Bishop, Matthew, and Michael Green. *Philanthrocapitalism: How the Rich Can Save the World*. New York: Bloomsbury Press, 2008.

Blumm, Michael C. *Sacrificing the Salmon: A Legal and Policy History of the Decline of Columbia Basin Salmon*. Lake Mary, FL: Vandeplas Publishing, 2013.

Bosso, Christopher J. *Environment, Inc.: From Grassroots to Beltway*. Lawrence: University Press of Kansas, 2005.

Brand, Stewart. *Clock of the Long Now: Time and Responsibility: The Ideas behind the Slowest Computer*. New York: Basic Books, 1999.

Brand, Stewart. *Whole Earth Discipline: An Ecopragmatist Manifesto*. New York: Viking Penguin, 2009.

Braungart, Michael, and William McDonough. *Cradle to Cradle: Remaking the Way We Make Things*. New York: North Point Press, 2002.

Braungart, Michael, and William McDonough. *The Upcycle: Beyond Sustainability—Design for Abundance*. New York: North Point Press, 2013.

Chouinard, Yvon. *Let My People Go Surfing: The Education of a Reluctant Businessman*. New York: The Penguin Press, 2005.

Christenson, Clayton. *The Innovator's Dilemma: The Revolutionary Book That Will Change the Way You Do Business*. New York: HarperCollins Publishers, 2000.

219

Christian-Smith, Juliet, and Peter H. Gleick with Heather Cooley, Lucy Allen, Amy Vanderwarker, and Kate A. Berry. *A Twenty-First Century U.S. Water Policy*. Oxford: Oxford University Press, 2012.

Collins, Jim. *Good to Great and the Social Sectors: Why Business Thinking Is Not the Answer.* New York: HarperCollins, 2005.

Collins, Jim, and Jerry I. Porras. *Built to Last: Successful Habits of Visionary Companies*. New York: HarperCollins, 1994.

Conkin, Paul K. *A Revolution Down on the Farm: The Transformation of American Agriculture since 1929*. Lexington: The University Press of Kentucky, 2008.

Cronon, William. *Nature's Metropolis: Chicago and the Great West*. New York: W.W. Norton & Company, 1991.

Darwin, Charles. *The Voyage of the Beagle*. Sydney: Beagle Press, 2013.

Diamandis, Peter H., and Steven Kotler. *Abundance: The Future Is Better Than You Think*. New York: Free Press, 2012.

Diamandis, Peter H., and Steven Kotler. *Bold: How to Go Big, Create Wealth and Impact the World*. New York: Simon & Schuster, 2015.

Fishman, Charles. *The Big Thirst: The Secret Life and Turbulent Future of Water*. New York: Free Press, 2011.

Friedman, Thomas H. *Hot, Flat and Crowded: Why We Need a Green Revolution— and How It Can Renew America*. New York: Picador, 2009.

Gleick, Peter H. *Water in Crisis: A Guide to the World's Fresh Water Resources*. Oxford: Oxford University Press, 1993.

Glennon, Robert. *Water Follies: Groundwater Pumping and the Fate of America's Fresh Waters*. Washington, DC: Island Press, 2002.

Glennon, Robert. *Unquenchable: America's Water Crisis and What to Do about It*. Washington, DC: Island Press, 2009.

Goldberg, Steven H. *Billions of Drops in Millions of Buckets: Why Philanthropy Doesn't Advance Social Progress*. Hoboken, NJ: John Wiley & Sons, 2009.

Hardin, Garrett. *Exploring New Ethics for Survival: The Voyage of the Spaceship Beagle*. New York: Penguin, 1973.

Hawken, Paul. *Growing a Business*. New York: Fireside, 1988.

Hawken, Paul, Amory Lovins, and L. Hunter Lovins. *Natural Capitalism: Creating the Next Industrial Revolution*. Boston: Little, Brown and Company, 1999.

Heinberg, Richard. *The End of Growth: Adapting to Our New Economic Reality*. Gabriola Island, BC: New Society Publishers, 2011.

Isaacson, Walter. *The Innovators: How a Group of Hackers, Geniuses, and Geeks Created the Digital Revolution*. New York: Simon & Schuster, 2014.

Jackson, Tim. *Prosperity without Growth: Economics for a New Planet*. New York: Earthscan, 2009.

Kahrl, William L. *Water and Power: The Conflict over Los Angeles Water Supply in the Owens Valley*. Berkeley: University of California Press, 1982.

Kawasaki, Guy. *The Art of the Start: The Time-Tested, Battle-Hardened Guide for Anyone Starting Anything*. New York: Penguin Group, 2004.

Leopold, Aldo. *Round River*. Oxford: Oxford University Press, 1993.

Leopold, Aldo. *A Sand County Almanac: With Essays on Conservation from Round River*. Oxford: Oxford University Press, 1966.

Lichatowich, Jim. *Salmon without Rivers: A History of the Pacific Salmon Crisis*. Washington, DC: Island Press, 1999.

Lynas, Mark. *The God Species: Saving the Planet in the Age of Humans*. Washington, DC: National Geographic Society, 2011.

McPhee, John. *Encounters with the Archdruid*. New York: Farrar, Straus and Giroux, 1980.

Midkiff, Ken. *Not a Drop to Drink: America's Water Crisis (and What You Can Do about It)*. Novato, CA: New World Library, 2007.

Pearce, Fred. *When the Rivers Run Dry: Water—The Defining Crisis of the Twenty-First Century*. Boston: Beacon Press, 2006.

Pollan, Michael. *The Omnivore's Dilemma: A Natural History of Four Meals*. New York: The Penguin Press, 2006.

Powell, John Wesley. *The Exploration of the Colorado River and Its Canyons*. New York: Penguin Classics, 2003.

Powell, John Wesley. *Report on the Lands of the Arid Region of the United States*. Cambridge, MA: Harvard Common Press, 1983.

Prud'homme, Alex. *The Ripple Effect: The Fate of Freshwater in the Twenty-First Century*. New York: Scribner, 2011.

Reisner, Marc. *Cadillac Desert: The American West and Its Disappearing Water*, revised edition. New York: Penguin Books, 1993.

Senge, Peter M. *The Fifth Discipline: The Art & Practice of the Learning Organization*. New York: Currency Doubleday, 1990.

Seuss, Dr. *The Lorax*. New York: Random House, 1971.

Shabecoff, Philip. *A Fierce Green Fire: The American Environmental Movement*. Washington, DC: Island Press, 2003.

Speth, James Gustave. *Angels by the River: A Memoir*. White River, VT: Chelsea Green Publishing, 2014.

Speth, James Gustave. *The Bridge at the Edge of the World: Capitalism, the*

Environment, and Crossing from Crisis to Sustainability. New Haven, CT: Yale University Press, 2008.

Stegner, Wallace. *Beyond the 100th Meridian: John Wesley Powell and the Second Opening of the West.* New York: Penguin Books, 1992.

Suzuki, David, and Ian Hanington. *Everything under the Sun: Toward a Brighter Future for a Small Planet.* Vancouver, BC: Greystone Books, 2012.

Tercek, Mark R., and Jonathan S. Adams. *Nature's Fortune: How Business and Society Thrive by Investing in Nature.* New York: Basic Books, 2013.

Wallace, David Rains. *The Klamath Knot.* San Francisco: Sierra Club Books, 1984.

Workman, James G. *Heart of Dryness: How the Last Bushmen Can Help Us Endure the Coming Age of Permanent Drought.* New York: Walker Publishing Company, 2009.

Worster, Donald. *A River Running West: The Life of John Wesley Powell.* Oxford: Oxford University Press, 2001.

Index

Page numbers followed by "f" and "t" indicate figures and tables.

Island Press | Board of Directors

Katie Dolan
(Chair)
Environmental Writer

Pamela B. Murphy
(Vice-Chair)

Merloyd Ludington Lawrence
(Secretary)
Merloyd Lawrence, Inc.
and Perseus Books

William H. Meadows
(Treasurer)
Counselor and Past President
The Wilderness Society

Decker Anstrom
Board of Directors
Discovery Communications

Stephen Badger
Board Member
Mars, Inc.

Terry Gamble Boyer
Author

Paula A. Daniels
Founder
LA Food Policy Council

Melissa Shackleton Dann
Managing Director
Endurance Consulting

Margot Paul Ernst

Tony Everett

Lisa A. Hook
President and CEO
Neustar Inc.

Mary James
Executive Committee Member
Prime Group, LLC

Charles C. Savitt
President
Island Press

Alison Sant
Cofounder and Partner
Studio for Urban Projects

Ron Sims
Former Deputy Secretary
US Department of Housing
and Urban Development

Sarah Slusser
Principal
emPower Partners, LLC

Deborah Wiley
Chair
Wiley Foundation, Inc.